Fundamentals of Algebra

Problems with Step by Step Solutions

Marcel Sincraian, Ph.D.

ISBN: 978-1-7388040-3-0

Marcel Sincraian Email: msincraian@yahoo.ca

To
My wife and children

CONTENT

Author's Notes

This book provides students with a tool to improve their knowledge of Algebra in preparation to learning Calculus. This book is intended to be a quick review of the fundamental concepts of Algebra studied in high-school.

It starts with the Concepts in Algebra. From associative property introduction to powers, and logarithms to radicals and ordered pairs. Then it follows with Straight lines chapter. Linear relations, distance between points, to slope of a line. It follows with Linear Equations. One step, two steps linear equations to straight line graphs, horizontal and vertical lines to parallel and perpendicular lines. Next chapter deals with linear inequalities. Next are the polynomials and factoring polynomial expressions. Next chapter is about Functions. Linear and quadratic, inverse functions, piecewise and trigonometric functions, to logarithmic functions. The next deals with transformation of functions. The last chapter is about systems of equations.

Chapter 1

Concepts in Algebra

1.A. SIMPLIFYING CONCEPTS IN ALGEBRA

a. Associative, commutative and distributive properties

Associative property

In a mathematical expression, the order of operations doesn't matter as long as the operations used in that particular case are only addition or only multiplication.

EXAMPLE

$(5+3)+7=5+(3+7)$

$8+7=15$

$5+10=15$

$(2 \times 5) \times 9 = 2 \times (5 \times 9)$

$10 \times 9 = 90$

$2 \times 45 = 90$

Commutative property

In a mathematical expression, the order of the numbers doesn't matter as long as the operations used in that particular case are only addition or only multiplication.

EXAMPLE

$5+7+3=7+5+3=15$

$2 \times 9 \times 7 = 7 \times 2 \times 9$

$18 \times 7 = 126$

$14 \times 9 = 126$

Distributive property

Whenever there is a bracket, the multiplication operation will distribute over either addition or subtraction operation that is inside that bracket.

EXAMPLE

$4 \times (5 + 6) = 4 \times 5 + 4 \times 6 = 20 + 24 = 44$

EXAMPLE

$3 \times (7 - 4) = 3 \times 7 - 3 \times 4 = 21 - 12 = 9$

Multiply:

$5 \times (x + 3)$

We apply the distributivity property and have:

$5 \times x + 5 \times 3 = 5x + 15$

Multiply:

$7x \times (x - 5) =$

We apply the distributivity property and have:

$7x \times (x - 5) = 7x \times x - 7x \times (-5) = 7x^2 - 35x$

Multiply:

$(3x + 4) \times 2x =$

We apply the distributivity property and have:

$(3x + 4) \times 2x = 3x \times 2x + 4 \times 2x = 6x^2 + 8x$

b. Expanding brackets, cross multiplication property

<u>Expanding brackets</u> means getting rid of them.

$a \times (b + c) = a \times b + a \times c$

EXAMPLE

$2 \times (7 + 4) = 2 \times 7 + 2 \times 4 = 14 + 8 = 22$

$6(2 + 5 - 2x) = 6 \times 2 + 6 \times 5 - 6 \times 2x = 12 + 30 - 12x = 42 - 12x$

FOIL (First, Outer, Inner, Last)

First, we multiply the first terms in each binomial. Then we multiply the Outer which means that we multiply the outermost terms in the product. Then Inner terms and Outer ones.

EXAMPLE

$(a + b)(c + d) = ac + ad + bc + bd$

We multiply the <u>First</u> terms $\boldsymbol{a} \times \boldsymbol{c}$

then we multiply the Outer terms $a \times d$,

We multiply the Inner terms $b \times c$,

and then we multiply the Last terms $b \times d$

Cross-multiplication means that when two fractions are equal the product between the numerator of first fraction and the denominator of the second fraction equals the product between the numerator of second fraction and the denominator of the first fraction.

If we have the fractions:

$\frac{a}{b} = \frac{c}{d}$, b and d are non-zero numbers

Then,

$a \times d = c \times b$

EXAMPLE

$\frac{4}{5} = \frac{12}{15}$

So, $4 \times 15 = 5 \times 12 = 60$

Application of cross multiplication.

Find x using cross multiplication property.

$\frac{6(x-3)}{2} = \frac{3(x+4)}{5}$

$5 \times 6(x - 3) = 2 \times 3(x + 4)$; $30x - 90 = 6x + 24$; $30x - 6x = 90 + 24$

$24x = 114$

$x = \frac{114}{24} = \frac{57}{12} = 4\frac{9}{12} = 4\frac{3}{4}$

1.B. INTRODUCTION TO POWERS

a. Definition, rules

The <u>power</u> is the group of two numbers where one is the base and the other is the exponent. The exponent tells us how many times we multiply the base with itself again and again.

EXAMPLE

The power, in this case, is the group 2^4

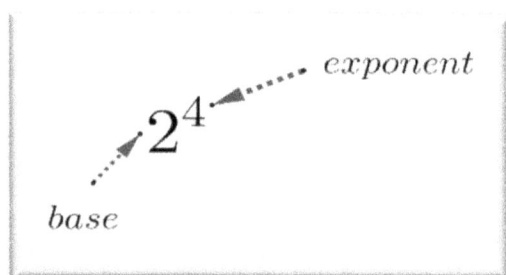

The base is 2, and the exponent is 4. This means that, the base 2 is multiplied with itself 4 times.

$2 \times 2 \times 2 \times 2 = 16$

EXAMPLE

Write the following repeated multiplications as a power.

$3 \times 3 \times 3 \times 3 \times 3 \times 3 = 3^6$

$5 \times 5 \times 5 \times 5 = 5^4$

EXAMPLE

Write the following powers as repeated multiplications.

$7^5 = 7 \times 7 \times 7 \times 7 \times 7$

$8^3 = 8 \times 8 \times 8$

Power rules

We consider a, b as non-zero real numbers, and m, n integers.

1. The product of powers with the same base

$a^n \times a^m = a^{n+m}$

EXAMPLE

$$3^5 \times 3^7 = 3^{5+7} = 3^{12}$$

2. The division of powers with the same base

$$a^n \div a^m = a^{n-m}$$

EXAMPLE

$$5^8 \div 5^6 = 5^{8-6} = 5^2$$

3. The product of two numbers at an exponent.

$$(a \times b)^n = a^n \times b^n$$

EXAMPLE

$$(6 \times 3)^4 = 6^4 \times 3^4$$

4. The quotient of two different numbers at an exponent.

$$\left(\frac{a}{b}\right)^n = \frac{a^n}{b^n} \quad b \neq 0$$

EXAMPLE

$$\left(\frac{5}{3}\right)^4 = \frac{5^4}{3^4}$$

5. The negative exponent.

$$a^{-n} = \frac{1}{a^n} \quad a \neq 0$$

EXAMPLE

$$3^{-5} = \frac{1}{3^5}$$

6. Exponent at an exponent

$$(a^m)^n = a^{m \times n}$$

EXAMPLE

$$(2^3)^5 = 2^{3 \times 5} = 2^{15}$$

7. Exponent zero

$$a^0 = 1$$

EXAMPLE

$$(87654789)^0 = 1$$

8. Fractional exponents.

$$a^{\frac{m}{n}} = \sqrt[n]{a^m}$$

EXAMPLE

$$3^{\frac{2}{5}} = \sqrt[5]{3^2}$$

And now, put it all together.

EXAMPLE

$$\frac{3^5 \times 4^4 \times (2 \times 3)^6}{3^3 \times (\frac{4}{3})^2 \times 2^4} = \frac{3^5 \times 4^4 \times 2^6 \times 3^6}{3^3 \times \frac{4^2}{3^2} \times 2^4} = \frac{3^5 \times 4^4 \times 2^6 \times 3^6}{\frac{3^3}{3^2} \times 4^2 \times 2^4} = \frac{3^{5+6} \times 4^4 \times 2^6}{3^{3-2} \times 4^2 \times 2^4} = \frac{3^{11} \times 4^4 \times 2^6}{3^1 \times 4^2 \times 2^4} = \frac{3^{11}}{3^1} \times \frac{4^4}{4^2} \times \frac{2^6}{2^4} =$$

$$3^{11-1} \times 4^{4-2} \times 2^{6-4} = 3^{10} \times 4^2 \times 2^2 = 3^{10} \times (2^2)^2 \times 2^2 = 3^{10} \times 2^4 \times 2^2 = 3^{10} \times 2^{4+2} = 3^{10} \times 2^6$$

EXAMPLE

$$8^{\frac{1}{3}} \times 5^7 \times \frac{(75-4+256-56+2457)^0}{2^3 \times 5^6} = \sqrt[3]{8} \times 5^7 \times \frac{(2728)^0}{2^3 \times 5^6} = \frac{2 \times 5^7 \times 1}{2^3 \times 5^6} = \frac{2}{2^3} \times \frac{5^7}{5^6} = 2^{1-3} \times 5^{7-6} = 2^{-2} \times 5 =$$

$$\frac{5}{2^2} = \frac{5}{4}$$

PRACTICE

Determine which answer is correct

1) The simplified expression of $\frac{5x^6y^9}{15x^4y^3}$ is $\frac{1}{3}x^2y^6; x, y \neq 0$

2) The simplified expression of $(-3a^2b^3)^3$ is $-27a^6; a \neq 0$

3) The simplified expression of $(\frac{-7x^4y^3}{xy})^2(\frac{x^2yz}{7x^2z})^3 = (-7x^3y^2)^2(\frac{y}{7})^3 = 49x^6y^4 * \frac{y^3}{49*7} =$

$\frac{1}{7}x^6y^7; x, y \neq 0$

4) The result of the expression $(\frac{2}{5})^{-1} = \frac{5}{2} = 2.5$

5) The simplified expression of $(\frac{8x^{-5}}{24xy^{-3}})^{-1} = \frac{24xy^{-3}}{8x^{-5}} = 3\frac{x^6}{y^3} ; x, y \neq 0$

b. Radicals, rules

Remember that a number multiplied by itself twice is that number squared.

EXAMPLE

$2 \times 2 = 2^2 = 4$

The reverse operation from 4 to 2 is the square root.

$\sqrt{4} = 2$

Cube root of 8 is 2

$\sqrt[3]{8} = 2$

Square roots of a perfect square are rational numbers. Square roots of a non-perfect square are irrational numbers. Cubic roots of a perfect cubic number are rational numbers.

Cubic roots of a non-perfect cubic number are irrational numbers.

Remember that:

$\sqrt{4} = \sqrt{2} \times \sqrt{2} = \sqrt{2^2} = 2$

$\sqrt[3]{8} = \sqrt[3]{2} \times \sqrt[3]{2} \times \sqrt[3]{2} = \sqrt[3]{2^3} = 2$

So, $\sqrt[n]{x^n} = x$

Rules of radicals

a. $\sqrt[n]{a} \times \sqrt[n]{b} = \sqrt[n]{a \times b}$

EXAMPLE

$\sqrt[3]{5} \times \sqrt[3]{7} = \sqrt[3]{35}$

b. $\sqrt[n]{\frac{a}{b}} = \frac{\sqrt[n]{a}}{\sqrt[n]{b}} = \sqrt[n]{a} \div \sqrt[n]{b}$

EXAMPLE

$$\sqrt{\frac{16}{25}} = \frac{\sqrt{16}}{\sqrt{25}} = \frac{4}{5}$$

c. $\sqrt[3]{-27} = -3$

The odd root from negative numbers is a negative number.

The even root from negative numbers is an imaginary number.

EXAMPLE

$$\sqrt[3]{-27} = -3$$
$$\sqrt{-2} = \sqrt{-1 \times 2} = \sqrt{-1} \times \sqrt{2} = i\sqrt{2}$$

Here $\sqrt{-1} = i$ or *imaginary*

PRACTICE

1) The root of $\sqrt[4]{81x^4}$ is

2) The perimeter of a rectangle with length $L = 3\sqrt{20} + 5\sqrt{5}$ and width $W = 5\sqrt{20}$ is

$P =$

3) The simplified expression of $\dfrac{3\sqrt[3]{x^5y}}{27\sqrt{x^2y^4}}$ is

4) The numerical value of $\sqrt[3]{-81} \div \sqrt[3]{-3}$ is

5) The simplified form of $\dfrac{\sqrt{\sqrt{16}}}{\sqrt{\sqrt{625}}}$ is

c. Mixed radicals, conversions of radicals

When the number under the radical is not a perfect root, the result will be a mixed radical. <u>Mixed radical</u> is when we have a term (number and or letters) multiplied with a radical.

EXAMPLE

1. $\sqrt{180} = \sqrt{4 \times 9 \times 5} = \sqrt{4} \times \sqrt{9} \times \sqrt{5} = 2 \times 3 \times \sqrt{5} = 6\sqrt{5}$

$\sqrt{180}$ is called entire radical

$6\sqrt{5}$ is a mixed radical.

2. $\sqrt[3]{56x^4} = \sqrt[3]{8 \times 7 \times x^3 \times x} = \sqrt[3]{8} \times \sqrt[3]{7} \times \sqrt[3]{x^3} \times \sqrt[3]{x} = \sqrt[3]{2^3} \times \sqrt[3]{7} \times \sqrt[3]{x^3} \times \sqrt[3]{x} = 2x\sqrt[3]{7x}$

$\sqrt[3]{56x^4}$ is the entire radical

$2x\sqrt[3]{7x}$ is a mixed radical

Conversion of radicals

Sometimes, we need to convert the entire radical into a mixed radical. Sometimes we have to do the reverse, to transform a mixed radical into an entire radical.

<u>a. Entire radical to mixed radical</u>

EXAMPLE

Find the result. Write the result into a mixed radical

$\sqrt{112} = \sqrt{4^2 \times 7} = \sqrt{4^2} \times \sqrt{7} = 4\sqrt{7}$

$\sqrt{20x^5y^3} = \sqrt{4 \times 5 \times x^5 \times y^3} = \sqrt{4 \times x^4 \times x \times y^2 \times y \times 5} = \sqrt{4 \times x^4 \times y^2 \times x \times y \times 5} = \sqrt{4} \times \sqrt{x^4} \times \sqrt{y^2} \times \sqrt{5xy} = 2x^2y\sqrt{5xy}$

<u>b. Mixed radical into an entire radical</u>

EXAMPLE

Find the result. Write the result into an entire radical

$3\sqrt{5} = \sqrt{3^2 \times 5} = \sqrt{9 \times 5} = \sqrt{45}$

$5\sqrt[3]{7} = \sqrt[3]{5^3 \times 7} = \sqrt[3]{125 \times 7} = \sqrt[3]{875}$

$7xy^2\sqrt{3} = \sqrt{7^2 \times x^2 \times y^4 \times 3} = \sqrt{49 \times 3 \times x^2 \times y^4} = \sqrt{147x^2y^4}$

EXAMPLE

Find the result. Write the result into a mixed radical

$\sqrt{450} + \sqrt{882} - \sqrt{162} = \sqrt{45 \times 10} + \sqrt{2 \times 441} - \sqrt{2 \times 81} = \sqrt{9 \times 5 \times 2 \times 5} + \sqrt{2 \times 49 \times 9} - \sqrt{2 \times 81} = \sqrt{9 \times 25 \times 2} + \sqrt{2 \times 49 \times 9} - 9\sqrt{2} = \sqrt{9} \times \sqrt{25} \times \sqrt{2} + \sqrt{9} \times \sqrt{49} \times \sqrt{2} - \sqrt{81} \times \sqrt{2} = 3 \times 5 \times \sqrt{2} + 3 \times 7 \times \sqrt{2} - 9 \times \sqrt{2} = 15\sqrt{2} + 21\sqrt{2} - 9\sqrt{2} = 27\sqrt{2}$

PRACTICE

1) The mixed radical of $\sqrt{192}$ is

2) The mixed radical of $\sqrt{175}$ is

3) The mixed radical of $\sqrt{288}$ is

4) The entire radical of $8\sqrt{7}$ is

5) The entire radical of $6\sqrt{3}$ is

1.C. UNDERSTANDING THE TRANSITION FROM NUMBERS TO VARIABLES

Remember that in lower grades we learned that:

$5 + 7 = 12$

It was obvious that 5 plus 7 is 12.

Remember as well, that we had the problems where we were asked: what number plus 7 equals 12?

$\boxed{?} + 7 = 12$

Or what number minus 4 equals 5?

$\boxed{?} - 4 = 5$

In Algebra we substitute the unknown number with a letter, say x

So, the above relations will become:

$x + 7 = 12$

Or,

$x - 4 = 5$

X or any other letter is called the variable or the unknown.

Any expression and operations between expressions that include one or more variables, is part of algebra.

Equal sign concept

The equal sign is fundamental in mathematics. It tells us that whatever number, letter, expression is on the left side of the sign, is exactly the same as what is on the right side of the equal sign.

It is obvious that 5=5. But what is very important to realize here is that the equal sign tells us that the number 5 to the left of = is the same with number 5 to the right of = sign.

Now, if we have:

X+2=3

Here it means that the final value of x+2 has to be the same as 3. We will come back to this concept when we discuss equations and how to solve them.

1.D. MATHEMATICAL EXPRESSIONS - INTRODUCTION

Definition

Expressions are mathematical **statements** that contain at least two terms that have numbers or variables, or both, connected by an operator in between. The mathematical operator used can be of addition, subtraction, multiplication, or division.

Expressions with one term would be $3x, x, 7, where\ x\ is\ the\ variable\ or\ unknown$

Expressions with two terms would be: $2x - 3;\ x + 4;\ 7 + 5x$

When we substitute the variable with a number, the expression will have a certain result.

EXAMPLE

Substitute a value of n=2 into these expressions:

1) $3n + 5 =$ 2) $(n + 2)(n - 1) =$ 3) $\frac{n+6}{2} =$

Solution

    1) $3(2) + 5 = 6 + 5 = 11$

    2) $(2 + 2)(2 - 1) = 4 \times 1 = 4$

    3) $\frac{2+6}{2} = \frac{8}{2} = 4$

PRACTICE

Substitute a value of n=3 into these expressions and find the value of R:

1) $R = 3n$ 2) $R = 5n - 3$ 3) $R = 2 \times (3n + 1)$

Substitute a value of p=-4 into these expressions and find the value of R:

4) $R = (3p + 2)(p - 4)$

5) $R = 5 \times (p - 6) - 2p$

6) $R = 3 \times (-2p + 3)$

1.E. ORDERED PAIRS - INTRODUCTION

Remember when you learned in lower grades about the number line.

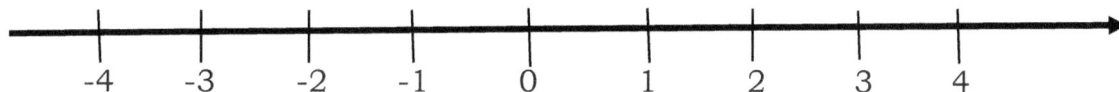

Each number is at a certain distance from the origin 0. Either to the left (negative numbers) or to the right (positive numbers).

In the 17th century Rene Descartes came up with the idea of using two number lines, one horizontal, and the other one vertical. From here, each point in the 2-dimensional plane received a pair of ordered coordinates, x for the horizontal line, and y for the vertical line. (x , y)
This is called the Cartesian System of axes.

EXAMPLE

If we want to represent any point situated anywhere in the 2-dimensional plane, we

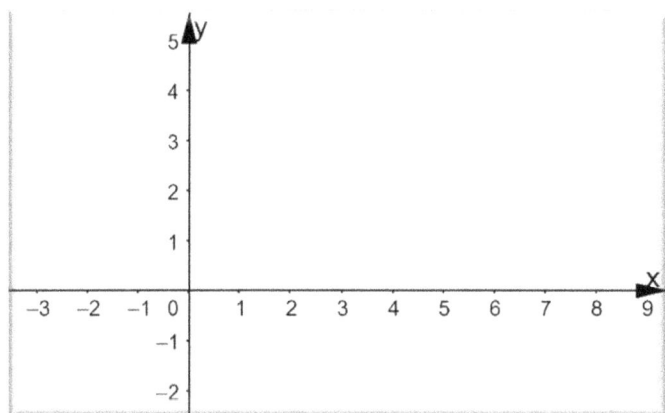

will have to give the point the coordinates x and y.
(x, y)
The intersection of the axes is called the system origin or center. The origin has the coordinates zero and zero. (0,0)

NOTE:
We always have the coordinates in the same order, first the x coordinate, then the y coordinate.
The system has four quadrants.
Quadrant 1 where both coordinates are positive.
Quadrant 2 where x is negative, and y is positive.
Quadrant 3 where both x and y coordinates are negative.
Quadrant 4 where x is positive, and y is negative.

Q2	Q1
Q3	Q4

Let's represent a few points in the Cartesian system.

EXAMPLE

Represent the following points.

A(2,3), B(-3,4), C(-2,-4), D(5,-3)

As it can be seen, for point A the x coordinate equals 2, and the y coordinate equals 3. For point D, for example, the x coordinate equals 5, and the y coordinate equals minus 3.

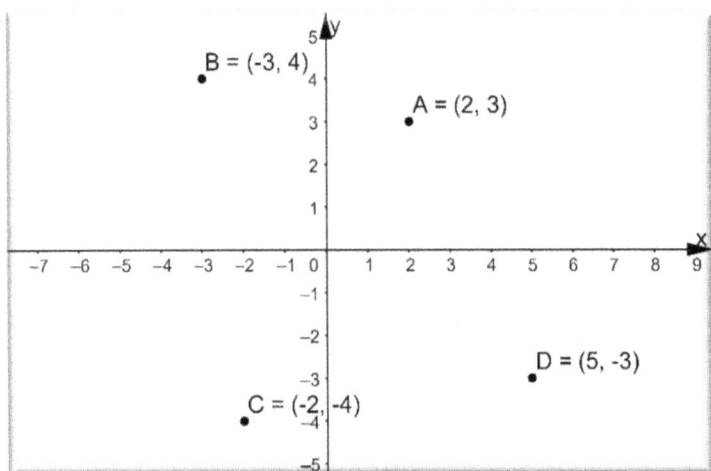

PRACTICE

Represent the following points on the graph below.

A (1,1)	B (3,5)
C (-2,4)	D (-3,1)
E (-3,-5)	F (-4,0)
G (4,-2)	H (3,-3)

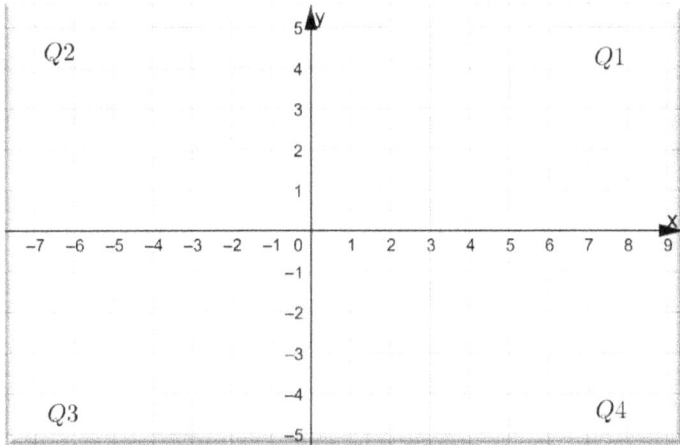

In what quadrants are the points below situated?

A (1,1.   B (3,5)

C (-2,4)   D (-3,1.

E (-3,-5)   F (-4,0)

G (4,-2)   H (3,-3)

Chapter 2

Straight Lines

2.A. REPRESENTING PATTERNS IN LINEAR RELATIONS

Let's suppose we want to count how many sticks we need to build 20 connected triangles.

We can see in the figure below that, in order to build one triangle, we need three sticks.

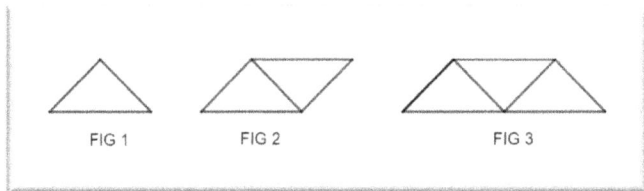

To build two connected triangles, we need 5 sticks. (FIG 2)
To build three connected triangles, we need 7 sticks. (FIG 3)

Let's gather all this information in a table with two columns.

Triangles	Sticks
1	3
2	5
3	7
4	9
20	41

We can see that as the number of triangles increases by 1, the number of sticks increases by 2 with each new connected triangle. If we continue this pattern, we will see that for 20 triangles we will need 41 sticks.

In the figure below, we have the case where we start with one square, then we add

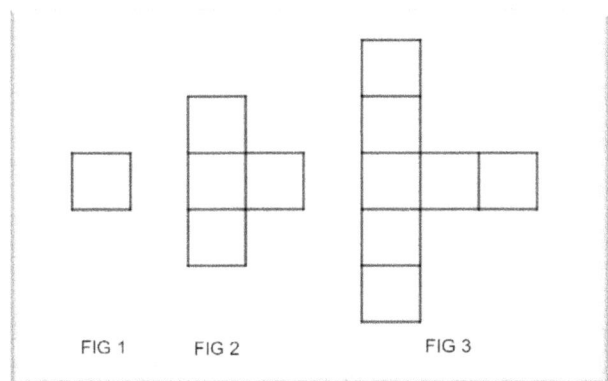

three squares, one above, one to the right and one below.
FIG 3 shows the case when we added another set of three squares to the ones in FIG 2.
How many squares would we need in FIG 7?

Again, we can gather all the information in a table with two columns. The first column contains the figure number. The second column contains the number of squares.

FIG #	Squares
1	1
2	4
3	7
4	10
7	19

We can see that as the FIG's number increases by 1, the number of squares increases by 3 with each new connected triangle. If we continue this pattern, we will see that in FIG 7 we will have 19 squares.

PRACTICE

1) Analyze the pattern shown in the figures below. Find how many houses figure 5 will have.

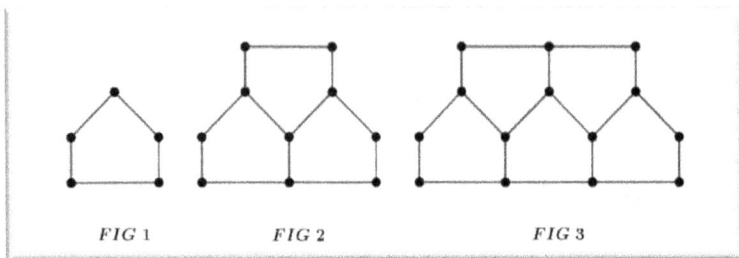

Figure	# Houses
1	1
2	3
3	5
5	

2) Analyze the pattern shown in figures below. Find how many squares figure 5 will have.

Figure	# Squares
1	2
2	4
3	6
4	8
5	

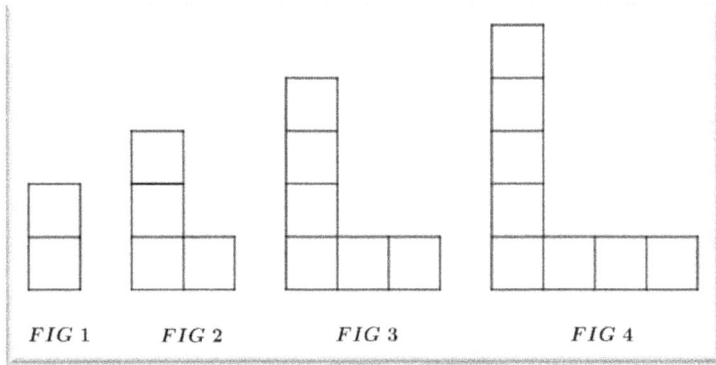

FIG 1   FIG 2   FIG 3   FIG 4

3) Analyze the pattern shown in figures below. Find how many triangles figure 5 will have.

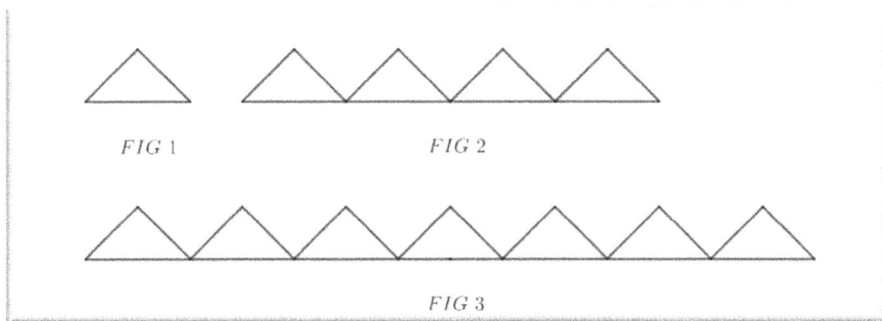

FIG 1   FIG 2

FIG 3

Figure	# Triangles
1	1
2	4
3	7
4	10
5	

4) Victoria needs to build cardboard cubes. After building the fourth cube, she is left with 31 carboard pieces. Can she finish the tenth cube?

Cube	# Carboard pieces
1	6
2	12
3	18
4	24

5) A rental car business has the client pay $40 for the first hour and $5 for every hour after that. How much does it cost to rent the car for 12 hours?

2.B. UNDERSTANDING TABLES OF VALUES OF LINEAR RELATIONSHIPS

Let's start with a table of values. The first column represents the x values. The second column represents the y values. Each <u>ordered</u> pair (x, y) represents a point in the system of axes.

Consider the table below.

X	Y
2	3
3	5
4	7
5	9

What do we notice in each column in the table above?

As successive x values increase by one unit, the successive y values increase with two units all the time.

For example, in the table above, in the x column, first value is 2, then 3, then 4, then 5. The difference between 3 and 2 is one, between 4 and 3 is again one, and so on. In the y column we have 3 as the first value, then 5, then 7, then 9. The difference between 5 and 3 is two. The difference between 7 and 5 is two again, and so on.

The increase in each column is the same with one unit for x and with two units for y respectively. The difference is not always 1 for x and 2 for y. It can be any difference, as long as it is the same for each successive value for x and y respectively.

EXAMPLE

Consider the table below:

X	Y
1	2
2	6
3	10
4	14

In the table above, in the x column, the first value is 1, then 2, then 3, then 4. The difference between 2 and 1 is one, between 3 and 2 is again one, and so on. In the y column we have 2 as the first value, then 6, then 10, then 14. The difference between 6 and 2 is four. The difference between 10 and 6 is four again, the difference between 14 and 10 is four again.

a. The relation between x and y in the tables

Consider the table below.

X	Y
2	3
3	5
4	7
5	9

As successive x values increase by one unit, the successive y values increase with two units all the time. For example, in the table above, in the x column, the first value is 2, then 3, then 4, then 5. The difference between 3 and 2 is one, between 4 and 3 is again one, and so on. In the y column we have 3 as the first value, then 5, then 7, then 9. The difference between 5 and 3 is two. The difference between 7 and 5 is two again, and so on. In the column of y we increase by 2 all the time

We can build a relation between x and y.

$y = 4x$

If we consider $x = 1$, then we have $y = 4 * 1 = 4$ which is two units more than 2.

So, to have $y = 4$ we need to subtract 2 to obtain $y = 2$

The general relation here will be $y = 4x - 2$

We can check with different values for x to see if the formula is good.

If $x = 4$ then we have $y = 4 * 4 - 2 = 16 - 2 = 14$, which is exactly the number we were expecting to get. The relation between y and x is indeed $y = 4x - 2$.

So, in order to create the relationship between y and x, we have to follow these steps: supposing that the difference between the two consecutive values in the x column is one:

Step 1

Check the difference between two consecutive values in the y column. (Let's call it A)

Step 2

Form the equation $y = A \times x$

Step 3

Check if for $x = 1$, the value for y at step 2 equals the value we should get in column y

Step 4

We add or subtract any value from $A \times x$ in such a way that we obtain the value of y that corresponds to the value of x in the x column.

Step 5

Check for another value of x to see if we obtain the correct corresponding value of y.

EXAMPLE

Consider the table below:

X	Y
1	7
2	10
3	13
4	16

Step 1

Check the difference between two consecutive values in the y column.

In this case that difference is 3

Step 2

Form the equation $y = 3 \times x$

Step 3

Check if for $x = 1$, the value for y at step 2 equals the value we should get in column y.

For $x = 1, y = 3 \times (1) = 3$

Step 4

We add or subtract any value from $A \times x$ in such a way that we obtain the value of y that is beside the value of x in the x column.

The value of y that corresponds to x=1 is 7 not 3.

We have to add 4 units to $y = 3 \times x$ in order to get 7.

The relation between y and x in this case will be:

$y = 3 \times x + 4$

Step 5

Check for another value of x to see if we obtain the correct corresponding value of y.

Let us take $x = 3$

Then we have: $y = 3 \times (3) + 4 = 9 + 4 = 13$ which equals the value of y for $x = 3$.

NOTE

If the first value of x is different than 1, we will take that value whichever it is.

PRACTICE

1) In the relation $M = 3k - 4$ determine M when k is:

a) 2   b) 17

c) 40   d) y+3

2) Determine the common difference in the pattern below.

$\sqrt{3} + \sqrt{5}, \quad \sqrt{3}, \quad \sqrt{3} - \sqrt{5} \ldots \ldots \ldots$

3) Analyze the table below and write a relation between x and y.

X	1	2	3	4	5
Y	15	10	5	0	-5

4) Determine the 20th term of the linear pattern below

5,9,13,17,21

We create the table:

X (term #)	1	2	3	4	5
Y	5	9	13	17	21

5) The total cost for a publishing company to publish a book is a fixed cost ($100) plus a cost for each additional book that the company will print. Create a general relation between the number of printed books and the cost of printing.

X (# of Books)	0	100	200	300	400
Cost	100	300	500	700	900

2.C. UNDERSTANDING GRAPHS OF LINEAR RELATIONSHIPS

Remember, the ordered pairs (x, y) represent points in the system of axes. If we represent the points from the table below,

Point	X	Y
A	1	2
B	2	5
C	3	8
D	4	11

we will have the graph below.

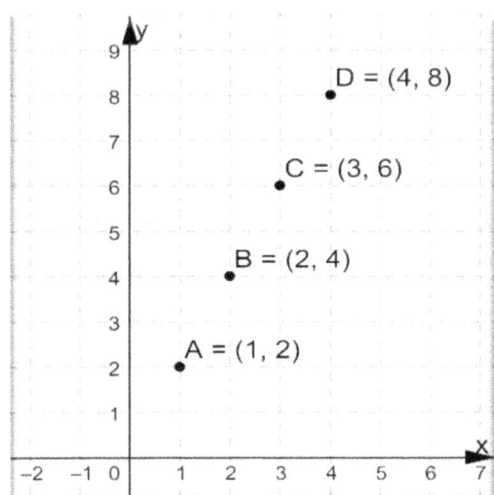

What we can see is that the points A, B, C, and D are situated on an imaginary straight line. If we connect these points, we will have a straight line.

PRACTICE

1) Represent graphically the points from the table below and see if they are part of a straight line.

Point	X	Y
A	1	-4
B	2	0
C	3	4
D	4	8

2) Represent graphically the points from the table below and see if they are part of a straight line.

Point	X	Y
A	1	-3
B	2	-1
C	3	1
D	4	3

3) Represent graphically the points from the table below and see if they are part of a straight line.

Point	X	Y
A	1	2
B	2	4
C	4	5
D	5	8

4) Represent graphically the points from the table below and see if they are part of a straight line.

Point	X	Y
A	1	1
B	2	3
C	3	5

5) Represent graphically the points from the table below and see if they are part of a straight line.

Point	X	Y
A	1	2
B	3	-3
C	5	6
D	4	7

2.D. THE DISTANCE BETWEEN POINTS

a. Horizontal distance

Let's suppose we have the points A (1,3) and B (5,3) As we can see the y coordinate is

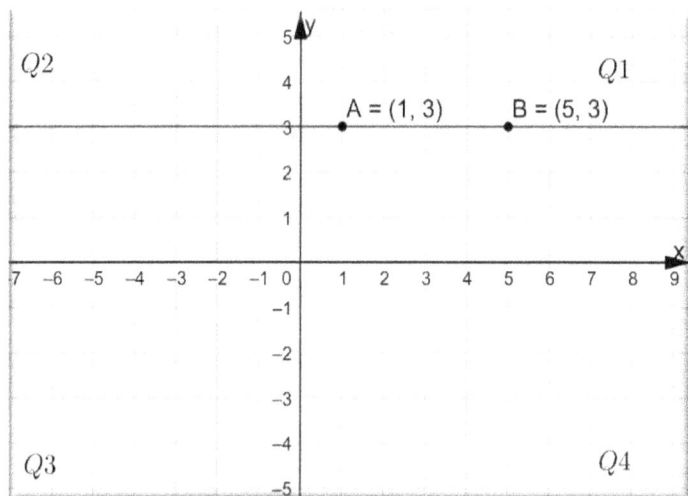

the same. If we represent these points on the cartesian system we get a horizontal segment that belongs to a horizontal line.

The distance between point A and point B will be 5 – 1 = 4 units.

Absolute value

Remember that the <u>absolute value</u> is always a positive number. We can think about the absolute value as the distance between the origin and that particular point, wherever the point is.

EXAMPLE

The absolute value of -7 is 7

The absolute value of -3.56 is 3.56

The notation for showing the absolute value is the number written between two vertical lines like below.

$|-7| = 7$ or $|-3.56| = 3.56$

If we go a bit more general, and consider Point A as point 1 and Point B as point 2, the coordinates of these two points will be written as:

A (x_1, y_1) and B (x_2, y_2)

The horizontal distance between A and B will be $|x_2 - x_1| = |5 - 1| = 4$

$|x_2 - x_1|$ represents the absolute value of the difference between the x coordinates of the points.

What happens when one of the x coordinates is negative?

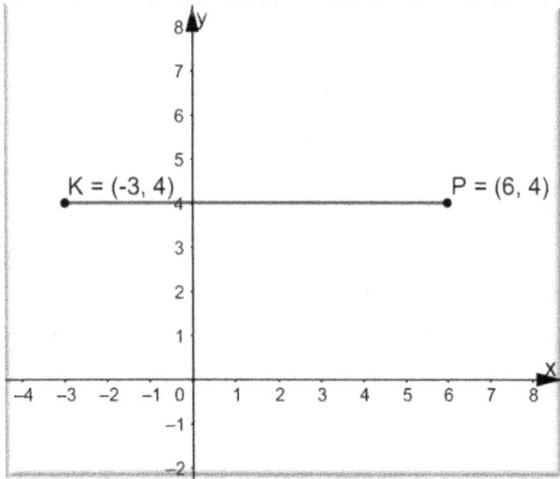

EXAMPLE

Let's calculate the horizontal distance between points K (-3,4) and P (6, 4)

K (x_1, y_1) and P (x_2, y_2)

The horizontal distance between K and P will be $|x_2 - x_1| = |6 - (-3)| = 9$

PRACTICE

Determine which answer is correct.

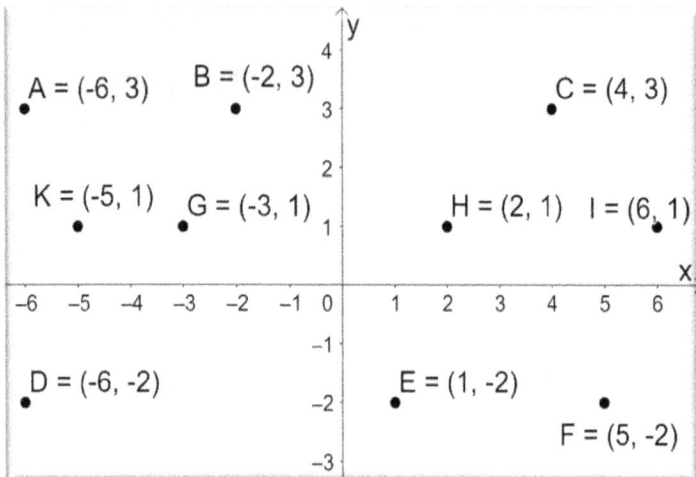

1) The distance between point A and point B is 5

2) The distance between point E and point F is 7

3) The distance between point K and point H is 7

4) The distance between point G and point H is 5

5) The distance between point H and point I is 4

b. *Vertical distance*

Let's suppose we have the points A (-3,1) and B (-3,6) As we can see the x coordinate is the same. If we represent these points on the cartesian system we get a vertical segment that belongs to a vertical line.

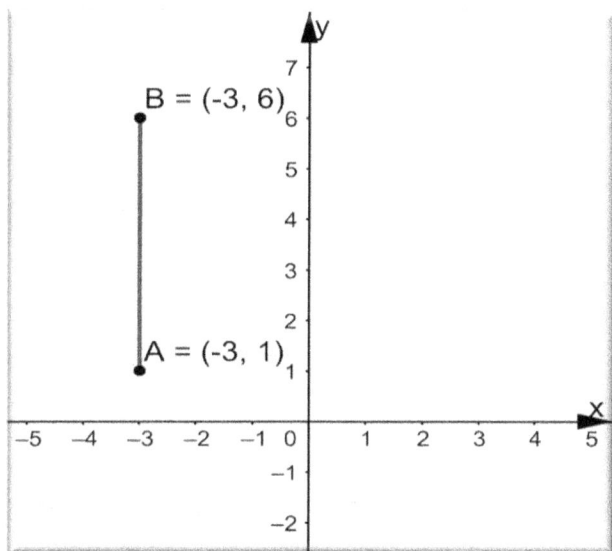

The vertical distance between point A and point B will be 6 – 1 = 5 units.

If we go a bit more general, and consider Point A as point 1 and Point B as point 2, the coordinates of these two points will be written as:

A (x_1, y_1) and B (x_2, y_2)

The vertical distance between A and B will be $|y_2 - y_1| = |6 - 1| = 5$

$|y_2 - y_1|$ represents the absolute value of the difference between the y coordinates of the points.

What happens when one of the y coordinates is negative?

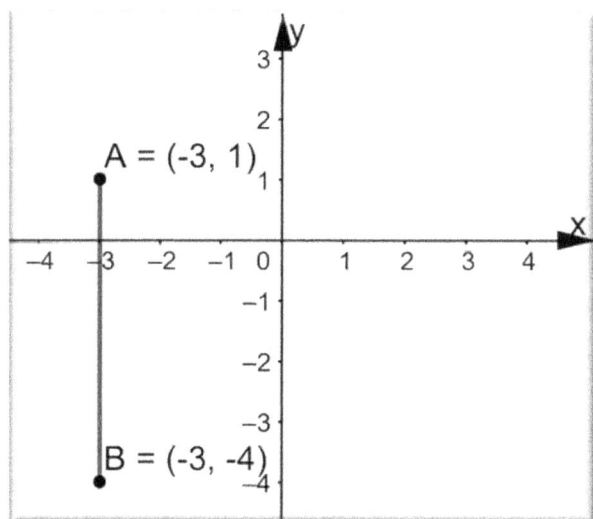

As we can see here, the y coordinates are 1 and -4 respectively.

The vertical distance between Point A and B will be:

$|y_2 - y_1| = |1 - (-4)| = 5$

PRACTICE

Determine if the relations below are true.

1) The distance between point A and point K is 10

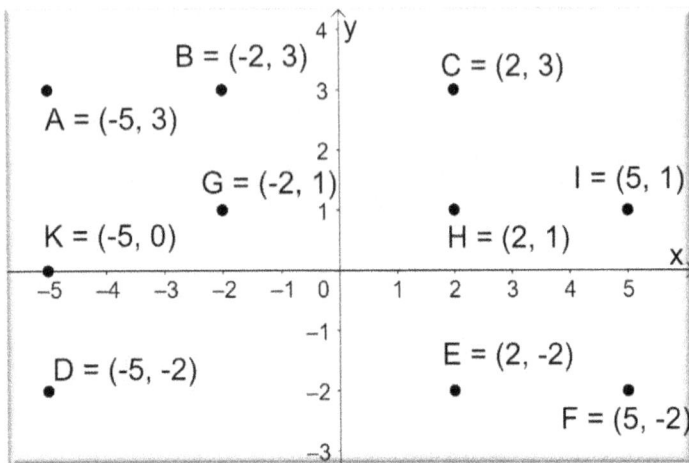

B = (-2, 3)   C = (2, 3)

A = (-5, 3)

G = (-2, 1)   I = (5, 1)

K = (-5, 0)   H = (2, 1)

D = (-5, -2)   E = (2, -2)

F = (5, -2)

2) The distance between point A and point D is 5

3) The distance between point B and point G is 3

4) The distance between point C and point H is 7

5) The distance between point C and point E is 5

6) The distance between point K and point D is 7

7) The distance between point H and point E is 3

8) The distance between point G and point E is 7

9) The distance between point A and point K is 9

10) The distance between point I and point H is 4

c. Non-horizontal and non-vertical distance

What happens when the points are in a straight line tilted from down left to up right?
What is the distance in this case?
We would like to calculate the distance between point A and B. The easiest way is to create a right-angle triangle as follows:

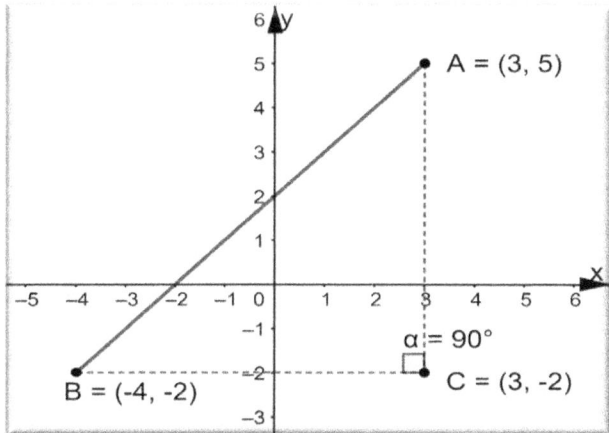

We draw a horizontal line from B to C that will intersect the vertical line from A to C.

The angle in the triangle ABC is a right-angle (90^0).

In this triangle, the distance between A and B represents the hypotenuse.

As we remember, the Pythagorean theorem tells us that in a right-angle triangle, the hypotenuse squared equals the sum of one side squared plus the other side of the triangle squared.

Here we have:

$$AB^2 = BC^2 + AC^2$$

BC is a horizontal segment; so, distance BC = $|x_2 - x_1| = |3 - (-4)| = |7| = 7$
AC is a vertical segment: so, distance AC = $|y_2 - y_1| = |5 - (-2)| = |7| = 7$

Then AB = $\sqrt{BC^2 + AC^2} = \sqrt{7^2 + 7^2} = \sqrt{98} = 9.8$

PRACTICE

Determine which answer is correct.

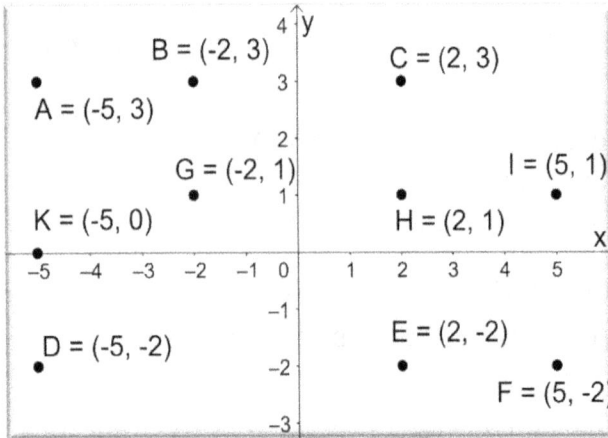

1) The distance between point B and point H is $2\sqrt{5} = 4.47$

2) The distance between point K and point C is 9.61

3) The distance between point A and point H is 7.28

4) The distance between point A and point F is 11.18

5) The distance between point B and point D is 10

6) The distance between point B and point F is 15

d. Midpoint coordinates

If we have two points $A(x_1, y_1)$ and $B(x_2, y_2)$, we need to find the coordinates of the point $C(x, y)$ situated at the middle distance between A and B.

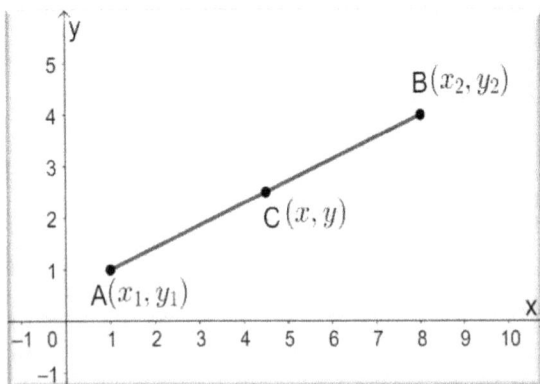

Because the distance between A and C equals the distance between C and B, to find the x coordinate, we can write:

$$x - x_1 = x_2 - x$$
$$2x = x_2 + x_1$$
$$x = \frac{x_2 + x_1}{2}$$

We do the same calculation on the y axis and have the formula of the y coordinate for the midpoint.

$$y - y_1 = y_2 - y$$
$$2y = y_2 + y_1$$
$$y = \frac{y_2 + y_1}{2}$$

EXAMPLE

Let's suppose the coordinates of A and B are shown in the figure below:

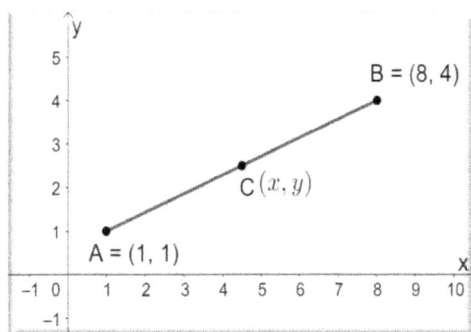

As we saw above, applying the formulas for the x and y coordinates of the midpoint C, we have:

$$x = \frac{x_2 + x_1}{2} = \frac{8+1}{2} = \frac{9}{2} = 4.5$$

$$y = \frac{y_2 + y_1}{2} = \frac{4+1}{2} = \frac{5}{2} = 2.5$$

So,

The midpoint coordinates are (4.5, 2.5

PRACTICE

Determine which answer is correct.

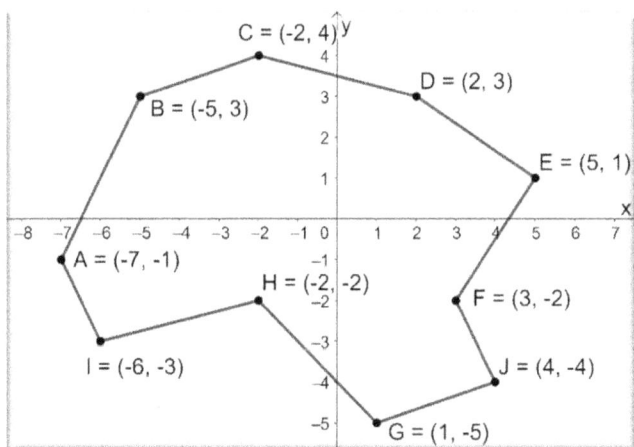

1) The mid-point coordinates of segment AB are: (-6,1)

2) The mid-point coordinates of segment BC are:(3,5)

3) The mid-point coordinates of segment CD are: (0,3.5)

4) The mid-point coordinates of segment DE are: (2,4)

5) The mid-point coordinates of segment EF are: (6,8)

6) The mid-point coordinates of segment FJ are: (3.5, -3)

7) The mid-point coordinates of segment JG are: (2.5, -4.5)

8) The mid-point coordinates of segment GH are: (0,-3)

9) The mid-point coordinates of segment HI are: (-4, -2.5)

10) The mid-point coordinates of segment IA are: (5, -3)

2.E. THE SLOPE OF A LINE

A slope of a line is the ratio between the "rise" of the line and the "run" of the line. The **"rise"** is the vertical distance or the difference between the y coordinates of any two points situated on the line

The **"run"** is the horizontal distance or the difference between the x coordinates of any two points situated on the line.

Let's suppose we want to find the slope of the line that passes through points P (x_1, y_1) and R (x_2, y_2).

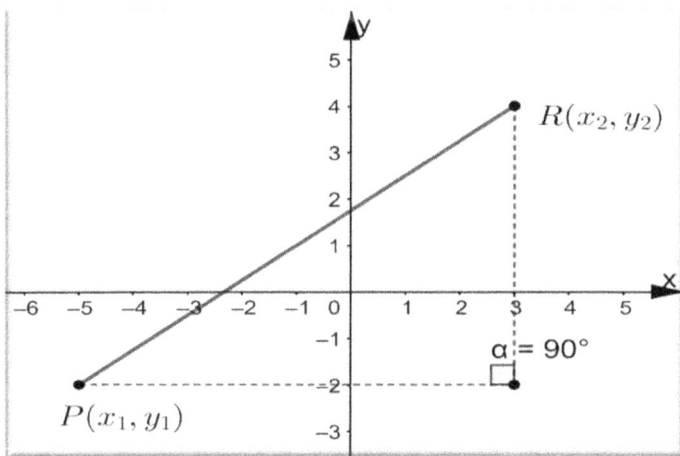

$$Slope = \frac{Rise}{Run} = \frac{Vertical\ distance}{Horizontal\ distance} = \frac{y_2 - y_1}{x_2 - x_1} = m$$

We could say that this ratio shows how fast the vertical distance increases as the horizontal distance increases.

EXAMPLE

We have a line that passes through points A (3,4) and B (-5,-2)

Let's suppose that B is point 2 and A is point 1

$$Slope = m = \frac{Rise}{Run} = \frac{Vertical\ distance}{Horizontal\ distance} = \frac{y_2 - y_1}{x_2 - x_1} = \frac{-2 - 4}{-5 - 3} = \frac{-6}{-8} = \frac{6}{8} = \frac{3}{4}$$

We could say that this ratio shows that the vertical distance increases by 3 units as the horizontal distance increases by 4 units.

Which will be the values of y as x increases from -5 to 3.

Using the relation for the slope that we just calculated, for $x_2 = -4$ we get:

$$\frac{3}{4} = \frac{y_2 - y_1}{x_2 - x_1} = \frac{y_2 - (-2)}{-4 - (-5)} = \frac{y_2 + 2}{1}$$

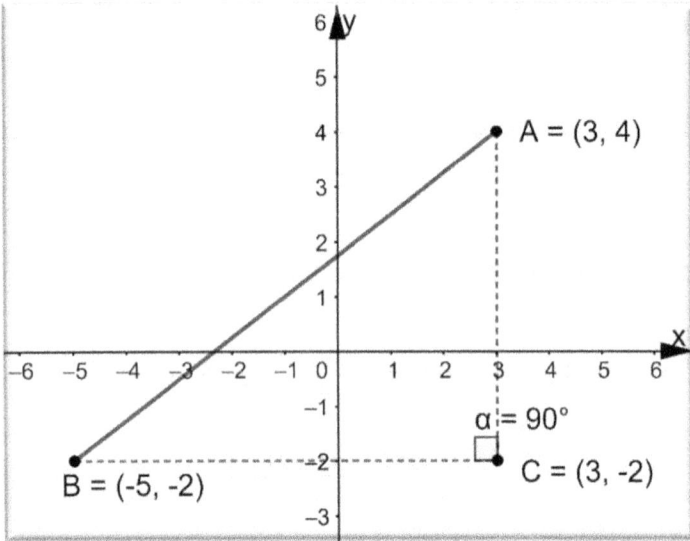

So,

$\frac{3}{4} = \frac{y_2+2}{1}$ using the cross-multiplication property

$3 = 4(y_2 + 2)$

$3 = 4y_2 + 8$

$3 - 8 = 4y_2$

$-5 = 4y_2$

$y_2 = \frac{-5}{4} = -1.25$

EXAMPLE

In the same way we can calculate the values for y for x increasing by 1 unit starting with   -5.

X	-5	-4	-3	-2	-1	0	1	2	3
Y	-2	-1.25	-0.5	0.25	1	1.75	2.5	3.25	4

As we can see, as the values of x increase, the values of y increase as well with the same ratio $\frac{3}{4}$.

Any increase (or change) of an unknown number or a variable (for example x) is symbolized Δx, where Δ is the Greek letter Delta.

So, the slope can be written as how fast the values of y vary as values of x vary.

$slope = \frac{\Delta y}{\Delta x}$ or could be thought of as the <u>average rate of change</u>.

PRACTICE

Determine if the relations below are true.

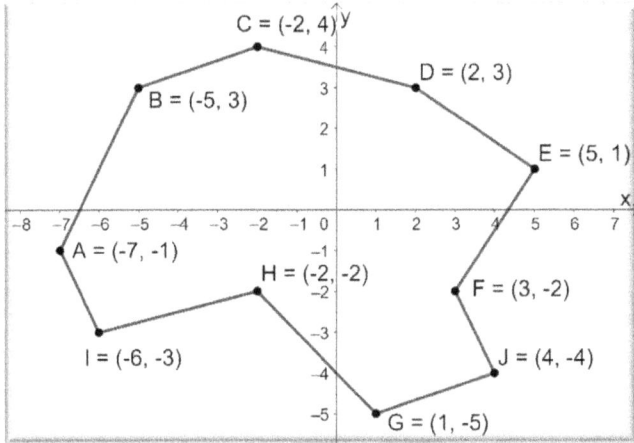

1) The slope of segment AB is: 2

2) The slope of segment BC is: 3

3) The slope of segment CD is: $-\frac{1}{4}$

4) The slope of segment DE is: 4

5) The slope of segment EF is: 5

6) The slope of segment FJ is: -2

7) The slope of segment JG is: $\frac{1}{3}$

8) The slope of segment GH is: 2

9) The slope of segment HI is: $\frac{1}{4}$

10) The slope of segment IA is: 7

Chapter 3

Linear Equations

3.A. LINEAR EQUATIONS

What is an equation?

An equation is a mathematical "statement" which says that the <u>mathematical expression</u> to the left of the equal sign is exactly the same as the <u>mathematical expression</u> to the left of the equal sign.

The expressions we discussed before are linear equations. $x - 4 = 5$ is an equation.

X is called a variable or unknown. This variable has the exponent 1. This is one of the reasons we call these relations linear.

The word equation comes from "equating" or regarding the sides of the equation as equal.

EXAMPLE

Let's find the unknown we symbolize with letter x in the following equation:

$x - 3 = 4$

The question is what number should we use for x, so that the left side of the equation equals the right side. In this case it is obvious that the number is 7. Here $x = 7$.

To find the unknown in an equation means to <u>solve</u> the equation.

3.B. SOLVING LINEAR EQUATIONS

a. Solve one-step linear equations:

First, we should talk about the balancing concept. What is this balancing concept? Whenever we have an equation, we could imagine that we have a scale that needs to be balanced all the time.

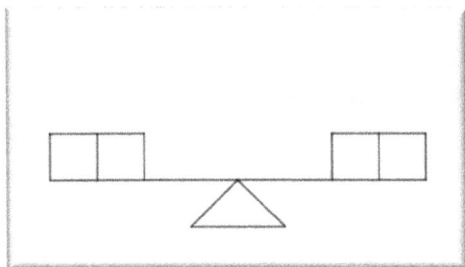

1. Let's suppose we have a balance (scale). On each side we have two identical boxes. The balance is completely horizontal.

2. Now, we want to put the third identical box to the left side of the balance. We need to keep the balance horizontal. In order to keep the balance horizontal, we need to put another box to the right side of the balance as well.

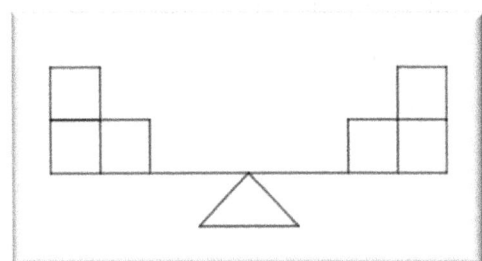

As we can see, we will add <u>exactly one</u> box on each side of the balance.

Think of the center of the balance as the equal sign of the equation. To keep this balance balanced or horizontal, we need to do the same action on each side of the balance.

3. Now, we want to take away two identical boxes from the left side of the balance. We need to keep the balance horizontal. In order to keep the balance horizontal, we need to take away another two identical boxes from the right side of the balance.

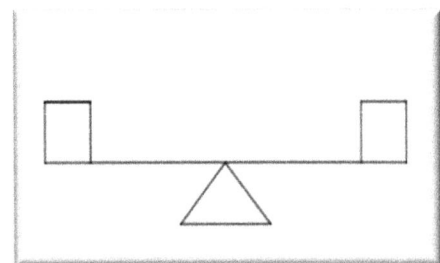

As we can see, we will take away <u>exactly two</u> boxes from each side of the balance.

EXAMPLE

1. We have the equation $x + 5 = 20$.
In order to <u>solve</u> the equation, or to find x, we need to <u>isolate</u> x.

- To do that we will need to cancel 5 from the left side.

- To cancel 5, we need to subtract the same 5 on the left side.

- To keep the equation balanced we have to subtract the same 5 on the right side.

3. To take away 5, we have;

$x + 5 = 20$

$\quad -5 \quad -5$

On the left side $+5 - 5 = 0$

Then,

$x + 0 = 20 - 5$

$x = 15$

We solved the equation by finding x.

To be sure that the result is correct, the next step is to <u>check</u> the solution.

EXAMPLE

We start with the previous equation:

$x + 5 = 20$

Where we found that $x = 15$

So,

We substitute 15 instead of x in the original equation.

The left side of the equation will be 15+5=20

The right side is equal with 20, so:

20=20

Now we can say for sure that $x = 15$ is the SOLUTION of the equation $x + 5 = 20$

PRACTICE

1) Solve and check the solution.

a) $x + 3 = 25$

b) $x + \frac{1}{2} = 3\frac{1}{3}$

2) Write each sentence as an equation

a) x decreased by 3 is 7

b) The sum of 4 and x is 9

3) Check if the given value of x is a solution for the equation.

a) $x + 3 = 7$ for $x = 5$

b) $-x - 2 = 7$ for $x = 4$

c) $x + 5 = 7$ for $x = 2$ d) $x + 3.4 = 5.8$ for $x = 4.3$

4) Explain the error.

$$x - \frac{2}{7} = \frac{5}{7}$$

$$x = \frac{3}{7}$$

5) The 30cm Length of a rectangle is bigger that the Width by 12cm. How much is the Perimeter?

b. Solving two-step linear equations with addition and subtraction

Now we can go one step further and analyze the case when we have the unknown in both sides of the equation.

EXAMPLE

$3x - 5 = 2x + 2$

We will have to ISOLATE the unknown.

We will do the same operations in both terms so the equation remains balanced all the time.

Step 1: minus $2x$ in both sides of the equation.

$3x - 2x - 5 = 2x - 2x + 2$

$x - 5 = 2$

Step 2: add a 5 on both sides

$x - 5 + 5 = 2 + 5$

$x = 7$

To be sure the result is correct, we check by substituting $x = 7$ in the original equation.

Check

$3(7) - 5 = 2(7) + 2$

$21 - 5 = 14 + 2$

$16 = 16$

Indeed, $x = 7$ is the <u>solution</u> of the equation $3x - 5 = 2x + 2$

EXAMPLE

$4x + 3 = 3x - 5$

We will have to ISOLATE the unknown.

We will do the same operations in both terms so the equation remains balanced all the time.

Step 1: minus $3x$ in both sides of the equation.

$4x - 3x + 3 = 3x - 3x - 5$

$x + 3 = -5$

Step 2: subtract a 3 on both sides

$x + 3 - 3 = -5 - 3$

$x = -8$

To be sure the result is correct, we check by substituting $x = -8$ in the original equation.

Check

$4(-8) + 3 = 3(-8) - 5$

$-32 + 3 = -24 - 5$

$-29 = -29$

Indeed, $x = -8$ is the <u>solution</u> of the equation $4x + 3 = 3x - 5$

PRACTICE

1) Solve and check

$2x - 4 = x + 2$

2) Solve and check.

$5x + \frac{1}{3} = 4x + 1\frac{1}{2}$

3) Solve ad check

$3.25x - 5 = 2.25x - 6$

4) Check if these given values of x are the solution for the equation below.

$3x - 5 = 5x + 3$ $\qquad x = 2, -3, -4$

5) Write an equation. Solve and check.

Four times a number increased by 3 is 3 times a number decreased by 1

$4x + 3 = 3x - 1$

c. Solving two-step linear equations with multiplication and division

This type of equations appears when we have a constant like 2 multiplied with the variable (x) after we isolated the variable.

EXAMPLE

$4x + 3 = 8$

In this case, to isolate the variable x, we have to subtract 3 and then divide both sides of the equation with the same value 4.

Here we divide by 4

$$\frac{4x}{4} = \frac{8}{4}$$

$x = 2$

EXAMPLE

$4x + 2 = 8$

We will have to ISOLATE the unknown or the variable.

We will do the same operations in both terms so the equation remains balanced all the time.

Step 1: minus 2 on both sides of the equation.

$4x + 2 - 2 = 8 - 2$

$4x = 6$

Step 2: divide by 4 on both sides

$$\frac{4x}{4} = \frac{6}{4}$$

$$x = \frac{6}{4} = \frac{3}{2}$$

To be sure the result is correct, we check by substituting $x = \frac{3}{2}$ in the original equation.

Check

$4(\frac{3}{2}) + 2 = 8$

$\frac{4*3}{2} + 2 = 8$

$\frac{12}{2} + 2 = 8$

$6+2=8$

$8=8$

Indeed, $x = \frac{3}{2}$ is the <u>solution</u> of the equation $4x + 2 = 8$

PRACTICE

1) Solve: $5x - 2 = 4$

2) Solve and check $4x + 3 = 4 + x$

3) Solve $\frac{5}{x} = 3$ $x \neq 0$

4) Solve and check $3.2x + 5.2 = 9.4 - 2.1x$

5) Fifteen divided by a number is 5. Write then solve an equation to determine the number. Verify the solution.

d. *Solving two- step linear equations using distributive property*

What is going to happen when we have the equation:

$2(x + 3) = 4$

We will first apply the distributivity property. We multiply the constant 2 with each of the terms in the bracket.

$2 * x + 2 * 3 = 4$

$2x + 6 = 4$

We subtract 6 from both sides.

$2x + 6 - 6 = 4 - 6$

$2x = -2$

We divide with 2 on each side.

$$\frac{2x}{2} = \frac{-2}{2}$$

$$x = -1$$

Check

We substitute $x = -1$ into the original equation to see if the right side equals the left side.

$$2[(-1) + 3] = 4$$

$$2(2) = 4$$

$$4 = 4$$

$x = -1$ is the solution of the equation $2(x + 3) = 4$

PRACTICE

1) Solve and check

$$3(x + 2) = 7$$

2) Solve and check

$$4(3x - 1) = 3(x + 5)$$

3) Solve and check

$$\frac{1}{2}(3x - 4) = \frac{3}{2}(2x + 5)$$

4) Solve

$$\frac{x}{3} + \frac{5}{3} = \frac{3}{4}$$

5) Solve

$$\frac{1}{3}(2x - 3) + 4x - 3 = \frac{5}{6}(x + 1) + 2$$

3.C. EQUATION OF A STRAIGHT LINE

a. Non-vertical and non-horizontal line

We have seen that the slope of a straight line is calculated with the formula:

$$Slope = m = \frac{Rise}{Run} = \frac{Vertical\ distance}{Horizontal\ distance} = \frac{y_2 - y_1}{x_2 - x_1}$$

So, if we have a point on a straight-line $M(x_1, y_1)$, then the slope of the line that passes through point M and any other point $K(x, y)$ will be written as:

$$m = \frac{y - y_1}{x - x_1}$$

So, if we isolate y from here, we have:

$y - y_1 = m(x - x_1)$ **(Called point-slope form)**

$y = y_1 + mx - mx_1$

So,

$y = mx + y_1 - mx_1$

But, $y_1 - mx_1$ is a constant (b)

So, we have the equation of a straight line

$y = mx + b$ **(Called slope-intercept form)**

Where:

m – slope

b – intersection of the line with y axis. It is called y-intercept

EXAMPLE

Find the equation of a line that passes through point M (2,3) and has a slope m=3

We start with the point slope equation;

$y - y_1 = m(x - x_1)$

We plug the coordinates of point M and the slope.

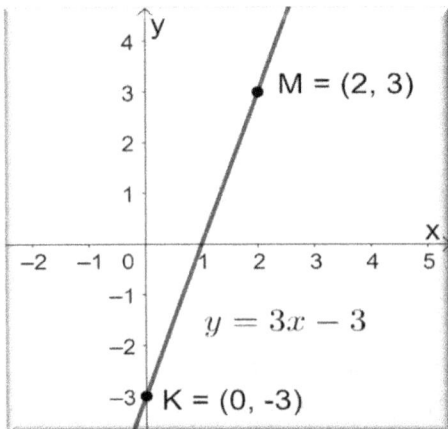

$$y - 3 = 3(x - 2)$$
$$y = 3 + 3x - 6$$
$$\mathbf{y = 3x - 3}$$

Here the slope is 3 and the intersection with the y axis is point K (0,-3)

Remember that the intersection of the line with x axis will have y coordinate equal to zero.

This is called x-intercept

$ax + bx + c = 0$ **(The general form of a linear equation)**

EXAMPLE

Represent the line $-8x + 4y - 4 = 0$

We transform the general form into the slope-intercept form in order to find the slope and the y intercept.

$$-8x + 4y - 4 = 0$$

Step 1: we add $8x$ on both sides of the equation.

$$-8x + 8x + 4y - 4 = 0 + 8x$$

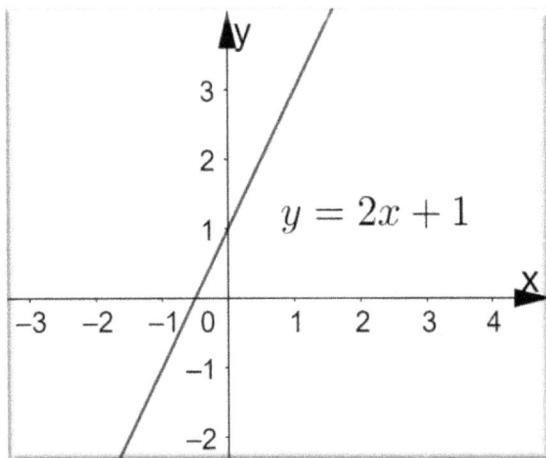

$$4y - 4 = 8x$$

Step 2: we add 4 on each side of the equation.

$$4y - 4 + 4 = 8x + 4$$
$$4y = 8x + 4$$

Step 3: we divide by 4 on each side of the equation.

$$\frac{4y}{4} = \frac{8x}{4} + \frac{4}{4}$$
$$y = 2x + 1$$

From here we notice that the slope of the line is 2 and the intersection with y axis or y intercept is 1.

b) Y-intercept and x-intercept

y-intercept is the point where the straight line intersects the y axis.

What does it mean in terms of the coordinates?

The y-intercept is a point situated on y axis. This means that its <u>x coordinate is zero.</u>

EXAMPLE

Represent the points A(0,5); B(0,3); C(0,-2); D(0,-4) on a cartesian system.

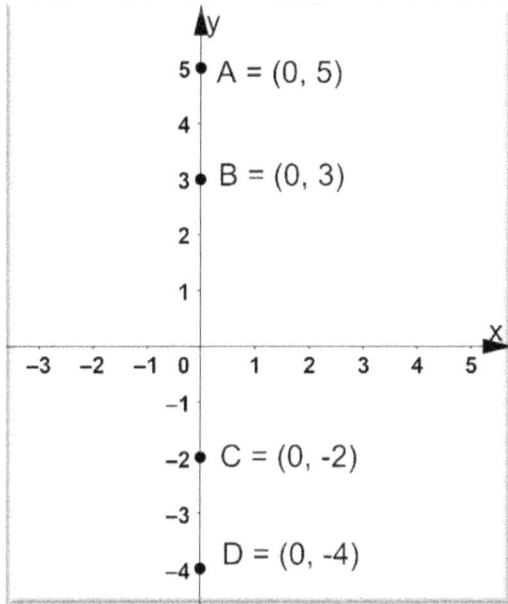

As we can see, all the points that are on the y axis have the x coordinate zero.

On the other hand, when we want to find the y intercept of a line, <u>we will write the condition x=0</u>

x-intercept is the point where the straight line intersects the x axis.

What does it mean in terms of the coordinates?

The x-intercept is a point situated on x axis. This means that its <u>y coordinate is zero.</u>

EXAMPLE

Represent the points A(4,0); B(3,0); C(-2,0); D(-3,0) on a cartesian system.

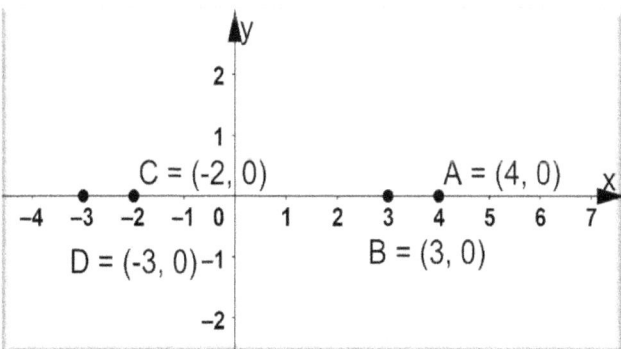

As we can see, all the points that are on the x axis have the y coordinate zero.

On the other hand, when we want to find the x intercept of a line, <u>we will write the condition y=0</u>

PRACTICE

1) The equation of the line through M (-3,1) and slope -2 is:

2) The y intercept of the line $y = -2x - 5$ is

3) In the slope relation, $m = \frac{y-5}{x+4}$, the y intercept in terms of the slope m, is

4) The equation of the parallel line with $y = 3x + 1$ that passes through the point M (5,6) is

5) The intersection to x axis of $y = 4x - 8\ is$ P (3,0)

3.D. STRAIGHT-LINE GRAPH

a. From equation to the graph

We have seen that the slope of a straight line is calculated with the formula:

$$Slope = m = \frac{Rise}{Run} = \frac{Vertical\ distance}{Horizontal\ distance} = \frac{y_2 - y_1}{x_2 - x_1}$$

So, if we have a point on a straight-line $M(x_1, y_1)$, then the slope of the line that passes through point M and any other point $K(x, y)$ will be written as:

$$m = \frac{y - y_1}{x - x_1}$$

So, if we isolate y from here, we have:

$$y = mx + b$$

Where:

m – slope

b – intersection of the line with y axis.

It is called y-intercept

EXAMPLE

Graph the straight-line represented by $y = 2x + 3$

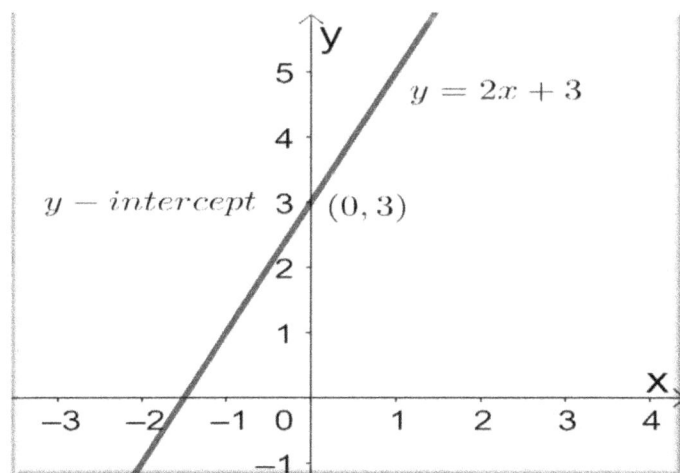

Here, the slope is 2, and the y intercept is the point (0,3)

EXAMPLE

Analyze each of these equations of straight lines and graph them.

a) $y = \frac{1}{2}x$

b) $y = -3x + 2$

c) $y = -x - 1$

d) $y = 3$

e) $x = -2$

Let's analyze them separately.

a) $y = \frac{1}{2}x$

the slope here is 0.5 and the y intercept is (0,0)

b) $y = -3x + 2$

Whenever the slope is negative, the line is tilted from down right to up left.
Here, the slope is -3 and the y intercept is (0,2).

c) $y = -x - 1$

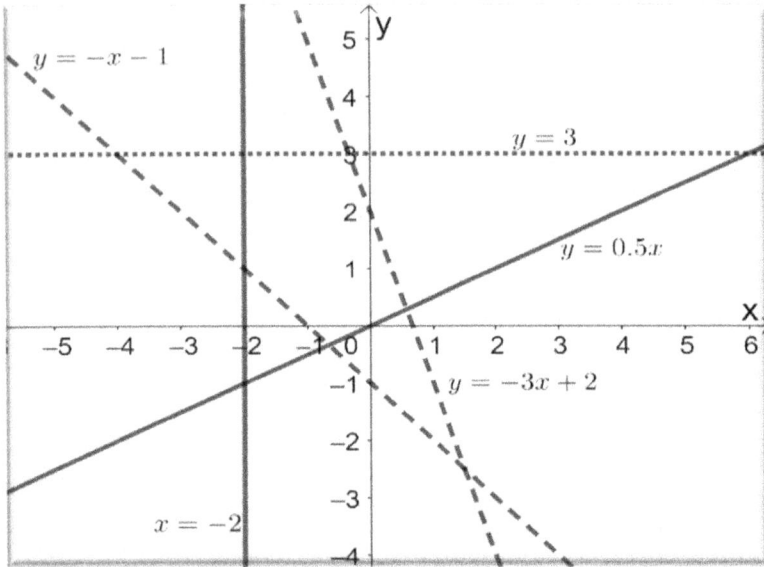

Here, the slope is negative (-1), so the line is tilted from down right to up left. The y intercept is (0,-1)

d) $y = 3$

Here, there is no x. This line is a horizontal line that goes through point (0,3) and is parallel with x axis.

e) $x = -2$

This is a vertical line that goes through point (-2,0) and is parallel with y axis.

b. From graph to the equation

Next, is the case when we are given the line, and we have to determine the mathematical equation of the line.

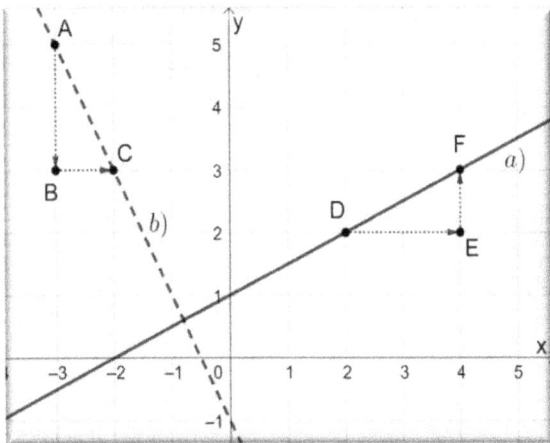

EXAMPLE

Find the equations of lines a) and b).

Line a)

We notice that y intercept is 1, so the constant in the equation is known.

To find the slope we start at point D, go horizontally (run) for +2 units to the right, then go vertically (rise) up plus 1 unit. So, the slope is calculated as:

$$slope = \frac{rise}{run} = \frac{1}{2}$$

The equation is $y = \frac{1}{2}x + 1$

Line b)

We notice that y intercept is -1, so the constant in the equation is known.

To find the slope we start at point A, go vertically ("rise" downward) for -2 units, then go horizontally (run) plus 1 unit. So, the slope is calculated as:

$$slope = \frac{rise}{run} = \frac{-2}{1} = -2$$

The equation is $y = -2x - 1$

PRACTICE

1) The slope of the line $y = 3x - 1$ is

2) The slope of the line $-3(x + 2) - 4(y - 7) = 6$ is

The next problems use the figure shown below

3) Line a) has the equation

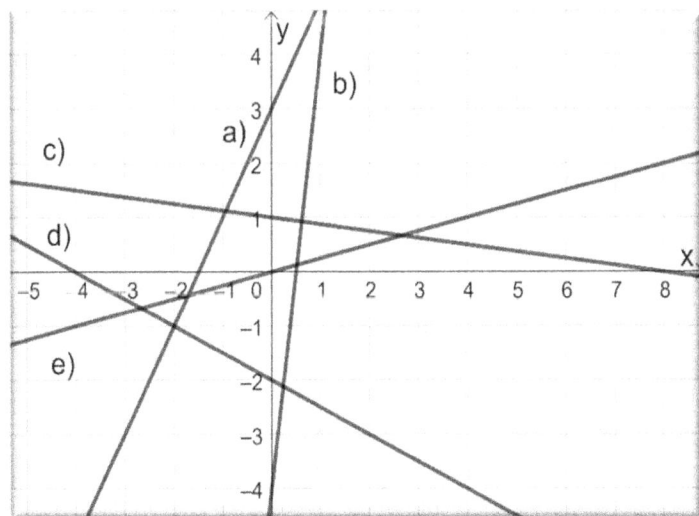

Y=

4) Line c) has the equation

$Y =$

5) Line e) has the equation

$y = \frac{1}{4}x$

3.E. SPECIAL CASES OF LINEAR EQUATIONS:

a. Vertical and horizontal lines

We have seen that the slope of a straight line is calculated with the formula:

$$Slope = m = \frac{Rise}{Run} = \frac{Vertical\ distance}{Horizontal\ distance} = \frac{y_2 - y_1}{x_2 - x_1}$$

So, if we have a point on a straight-line $M(x_1, y_1)$, then the slope of the line that passes through point M and any other point $K(x, y)$ will be written as:

$$m = \frac{y - y_1}{x - x_1}$$

Vertical line

Let's see an example of a vertical line. If we represent the coordinates of a few points on this vertical line, say x=3, we will notice that each point on this vertical line has the same x coordinate. This x coordinate is always 3. In this case the equation of a vertical line that passes through x equal 3 will have the equation

$x = 3$

We see that the equation for a vertical line is _x = constant_

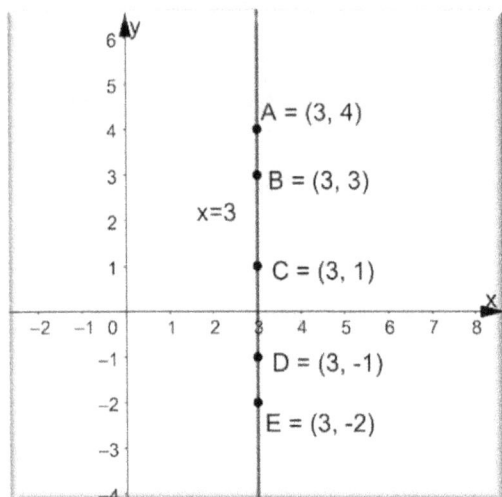

This means that for any values of y in the formula $m = \frac{y - y_1}{x - x_1}$, the x coordinates will always be the same. But, if the x is the same, the denominator of the slope formula will become zero. For the constant x, the slope m is infinite, or undefined.

Horizontal line

Let's see an example of a horizontal line. If we represent the coordinates of a few points on this horizontal line, say y=-2, we will notice that each point on this line has the same y coordinate. This coordinate is always minus 2. In this case the equation of a vertical line that passes through y equal minus 3 will have the equation

$y = -2$

The equation for the horizontal line is <u>$y = constant$</u>

In this case, the y coordinates of any point on the horizontal line are the same. From here we have that the difference $y - y_1$ will be zero.

For the constant y, the <u>slope m is equal to zero</u>.

EXAMPLE

Find the slope of the line with the equation y=3

Let's consider point one N (5,3)

and point two M (9,3) situated on the line.

Let's calculate the slope.

$m = \frac{y_2 - y_1}{x_2 - x_1} = \frac{3-3}{9-5} = \frac{0}{4} = 0$

PRACTICE

1) The equation of a horizontal line passing through M (3,4) is

2)

Next four problems will be based on the figure shown below.

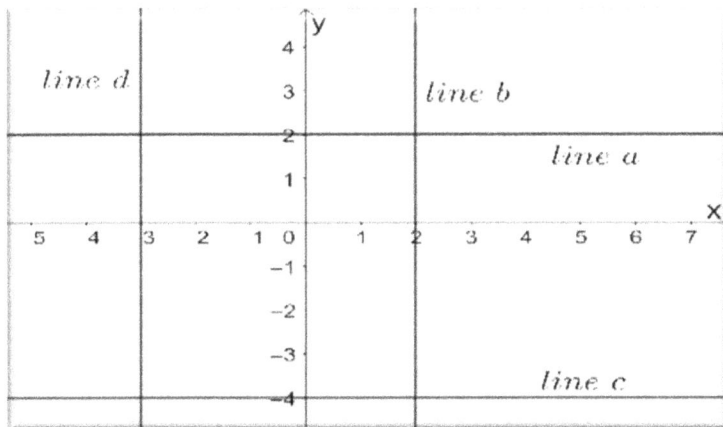

2) The equation of line a is

3) The equation of line b is

4) The equation of line c is

5) The equation of line d is

3.F. PARALLEL AND PERPENDICULAR LINES

a. Parallel lines

Remember that we have the formula for a straight line as:

$y = mx + b$

In this context, two lines are parallel when they have the same slope. The y intercept has to be different.

EXAMPLE

Graph the following straight lines.

$y = 2x + 1$

$y = 2x + 4$

As we can see in the graph, the equations have the same slope, in this case 2, and different y intercepts, in this case 1 and 4 respectively.

Question:

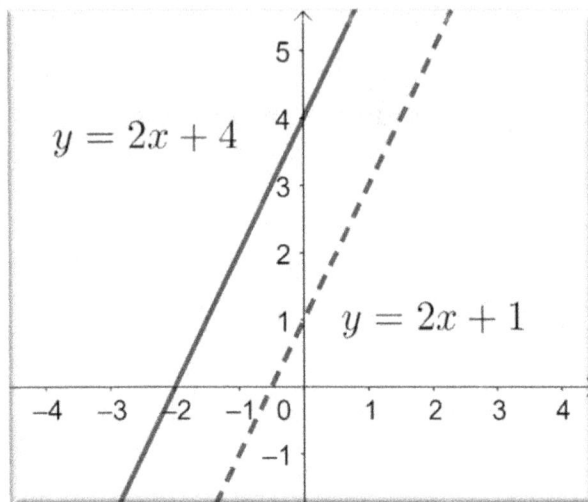

What happens when the slopes are the same, and the y intercepts are the same as well?

EXAMPLE

We have the lines.

$y_1 = 2x + 1$

$y_2 = 2x + 1$

We can see that these lines y_1 and y_2 are identical. Indeed, if we represent them on the cartesian system, it will be exactly the same line. We will have only one line.

PRACTICE

1) The equation of a line parallel with $y = x - 1$ that intersects y axis at point M (0,5) is

2) The line $y = -5x + 3$ is _____ parallel with $y = -4x + 3$

3) The line $y = 4x + 3$ is parallel with

Next problems will be based on the figure shown below.

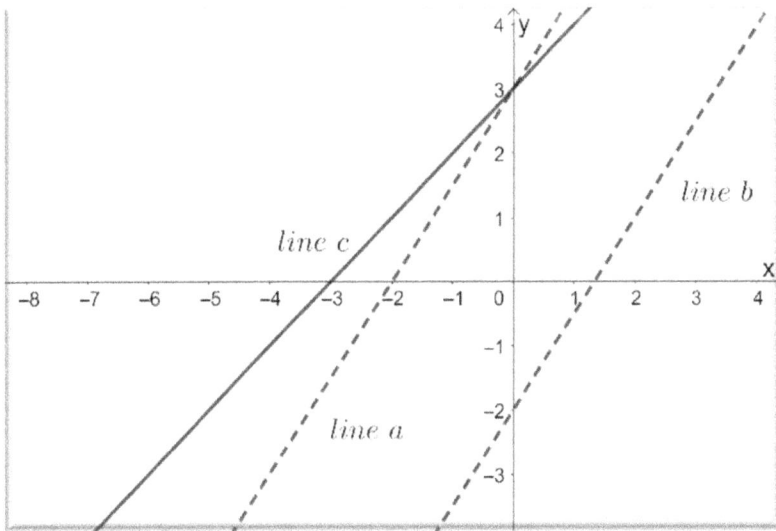

4) Line a is _____ to line b

5) The equation of the line b is

$y =$

b. *Perpendicular lines*

Remember that two lines that are not vertical are perpendiculars (i.e. have an angle of 90^0 between them) if the product of their slopes is negative 1.

This means that, if the first line's slope is m_1, and the second line's slope is m_2, then these lines are perpendicular if $m_1 \times m_2 = -1, or \ m_1 = \frac{-1}{m_2}$

EXAMPLE

We have the line given by the equation:

$y_1 = 2x + 1$

Show that the line given by the equation:

$y_2 = -0.5x + 4$

is perpendicular to the first line.

So,

$m_1 = 2, and\ m_2 = -0.5$

Then,

$m_1 \times m_2 = 2 \times (-0.5) = -1$

From here it results that y_1 and y_2 are perpendiculars.

These two lines are represented in the graph below.

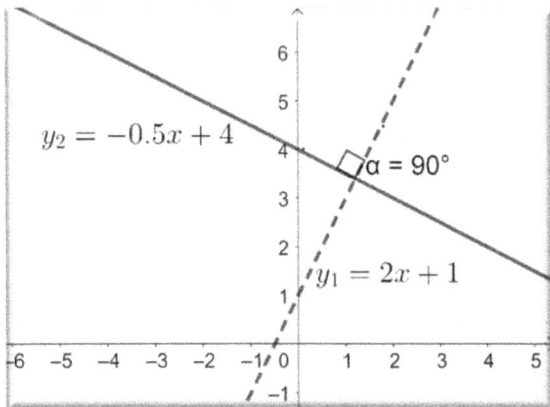

PRACTICE

1) The lines that have the equations $y = 3x + 4$ and $y = -\frac{1}{3}x - 5$ are _____.

2) The line perpendicular to the line $y = -2x + 7$ has the slope $m =$ _____.

3) The equation of the line perpendicular to $y = -2x + 3$ through M (-2,-3) is_____

4) The equation of the line perpendicular to AB in point B, is $y =$

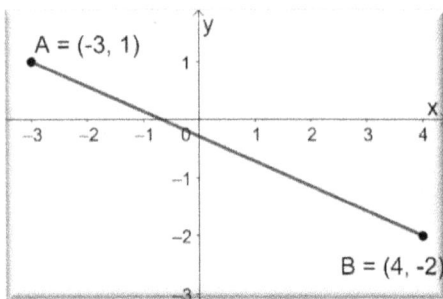

5) The perpendicular to AB through the point B (6,1) will intersect y axis in _____

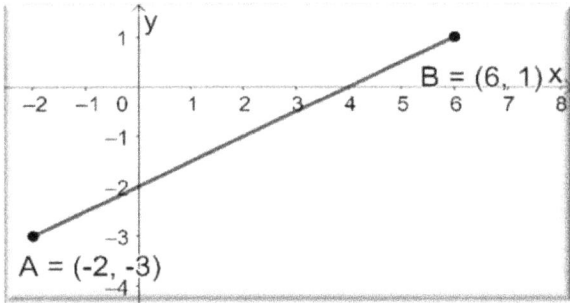

Chapter 4

Linear Inequalities

4.A. EXPRESS LINEAR INEQUALITIES GRAPHICALLY AND ALGEBRAICALLY

Let's start with a x value of, say 5. We represent it on the number line as a point at five units to the right of zero, or 5 units to the right of zero.

Algebraically, we write $x = 5$

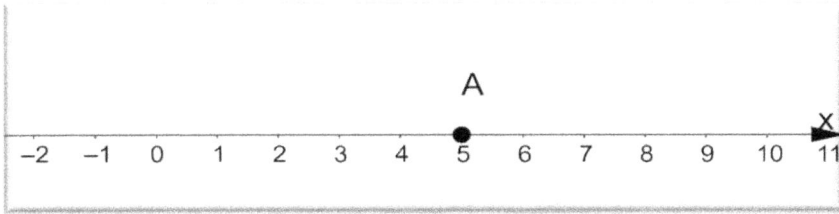

When instead, we refer to all the numbers smaller than, say 5, including 5 we represent this case as an arrow above the number line towards left, starting with a filled circle.

Algebraically we write: $x \leq 5$

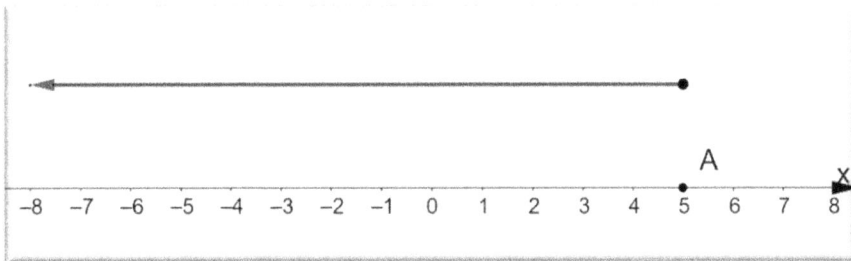

When we refer to all the numbers smaller than, say 3, excluding 3, we represent an arrow above the number line towards left, starting with an empty circle.

Algebraically we write: $x < 3$

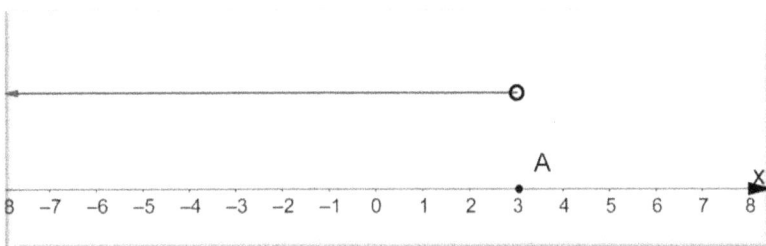

When we refer to all the numbers bigger than, say -3, including -3, we represent this case as an arrow above the number line towards right, starting with a filled circle.

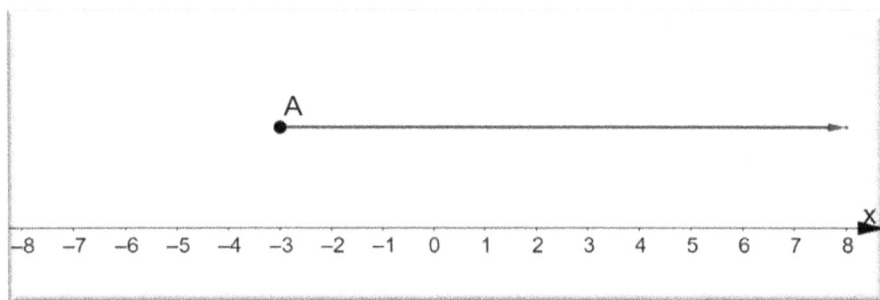

Algebraically we write:

$$x \geq -3$$

When we refer to all the numbers bigger than, say -7, excluding -7, we represent an arrow above the number line towards right, starting with an empty circle.

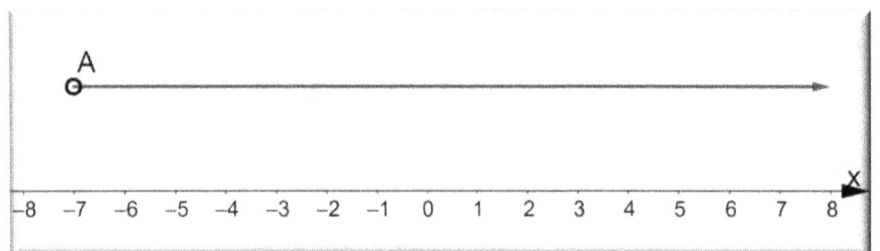

Algebraically we write:

$$x > -7$$

EXAMPLE

Represent on the line and algebraically:

a. a number bigger or equal to 0

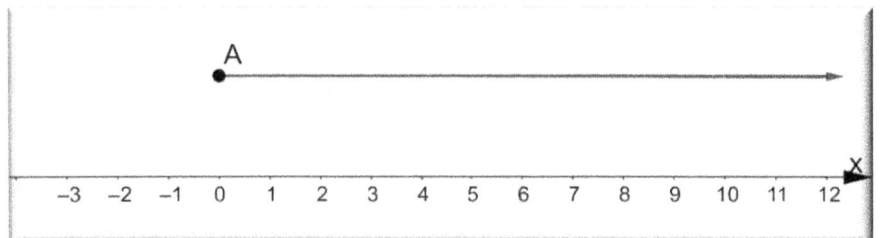

Algebraically we write:

$$x \geq 0$$

b. all numbers smaller and equal to 4

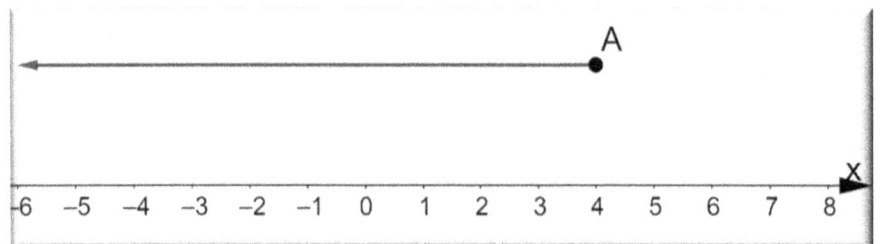

Algebraically we write:

$$x \leq 4$$

PRACTICE

1) Represent on the number line and algebraically:
A number bigger than and equal to -5

2) Represent on the number line and algebraically:
A number less than and equal to 2

3) Use a symbol $>, <, \leq, or \geq$ to write an inequality that corresponds to each statement.
a) x is less than -7
b) a number is greater and equal than 3
c) x is negative

4) Write 3 numbers that are solutions of each inequality.

a) $a > 3$ b) $b \leq 3$ c) $w < -5$

5) Iveta and Emma write the inequality whose solution is shown below.

A

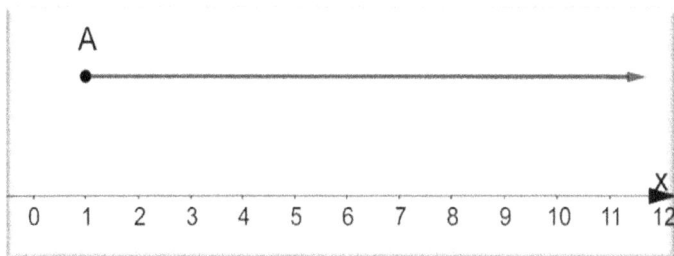

Iveta writes $x \geq 1$

Emma writes $1 > x$

Who is correct?

| 0 | 1 | 2 | 3 | 4 | 5 | 6 | 7 | 8 | 9 | 10 | 11 | 12 |

4.B. SOLVING ONE-STEP LINEAR INEQUALITIES

Let's have the inequality:

$x + 3 \leq 7$

Like in case of equations the idea is to isolate the unknown.

We minus 3 in both sides of the inequality.

$x + 3 - 3 \leq 7 - 3$

So,

$x \leq 4$

According to 3.1 we can represent this on number line.

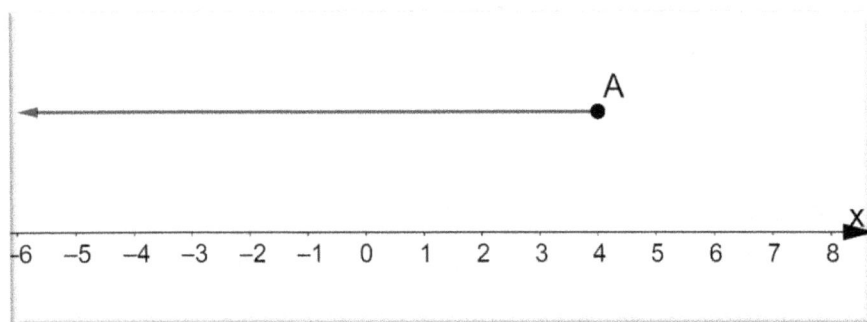

Let's have the inequality:

$x - 5 > 1$

Like in case of equations the idea is to isolate the unknown.

We add 5 in both sides of the inequality.

$x - 5 + 5 > 1 + 5$

So,

$x > 6$

According to 3.1, we can represent this solution on the number line.

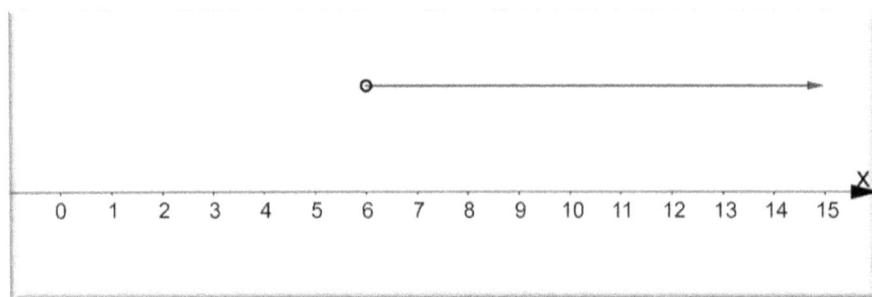

What would happen when we have:

$-x + 4 > 12$

We subtract 4 in both sides.

$-x + 4 - 4 > 12 - 4$

$-x > 8$

We want x not $-x$ so, we multiply with minus one on both sides of the inequality.

In this case, the sign of the inequality changes from $>$ to $<$

$x < -8$

EXAMPLE

Solve for x

$2x \geq 6$

We divide both sides with the coefficient of x.

$\frac{2x}{2} \geq \frac{6}{2}$

$x \geq 3$

PRACTICE

1) Solve and graph the solution on the number line.

$x - 7 \geq -4$

2) What must be done to the first inequality to get to the second inequality?

a) $x - 5 \leq 4$ b) $2x \leq 4$ c) $x - \frac{1}{3} > \frac{2}{3}$

$x \leq 9$ $x \leq 2$ $x > 1$

3) State three values that satisfy each inequality; one integer, one fraction, and one decimal

a) $x + 3 < 5$

b) $x - 4 > 1$

c) $5x \leq 20$

4) Melissa has $310 in her bank account. She must maintain a minimum balance of $600 in her bank account to avoid paying a monthly fee. How much money can Melissa deposit into her account to avoid paying bank fees?

a) Choose a variable and write an inequality to solve the problem.

b) Solve the problem

5) A water slide charges \$2 to rent an inflatable ring, and \$0.5 per ride. Iveta has \$12. How many rides can Iveta go on?

4.C. SOLVING MULTI-STEP LINEAR INEQUALITIES

Let's suppose we have:

$3x - 5 > 10$

First, we add 5 to each side to cancel -5 in the left side.

$3x - 5 + 5 > 10 + 5$

$3x > 15$

Second, we divide with the coefficient of the unknown.

$\frac{3x}{3} > \frac{15}{3}$

We get:

$x > 5$

EXAMPLE

$2x + 4x - 5 > 10 + 3$

Step 1: add 5 in both sides to cancel 5 in the left side.

$2x + 4x - 5 + 5 > 10 + 3 + 5$

Step 2: deal with the like terms in both sides.

$6x > 18$

Step 3: divide with the coefficient of the unknown in both sides.

$\frac{6x}{6} > \frac{18}{6}$

$x > 3$

EXAMPLE

$5x - 7 > 3x - 18 + 3$

Step 1: add 7 in both sides to cancel 7 in the left side.

$5x - 7 + 7 > 3x - 18 + 3 + 7$

$5x > 3x - 6$

Step 2: subtract 3x in both sides to isolate the unknown on the left side.

$5x - 3x > 3x - 3x - 8$

$2x > -8$

Step 3: divide with the coefficient of the unknown in both sides.

$\frac{2x}{2} > -\frac{8}{2}$

$x > 4$

EXAMPLE

$\frac{(x+2)}{2} < \frac{(3x-5)}{3}$

Step 1: multiply both sides with the common denominator 6

$\frac{6*(x+2)}{2} < \frac{6*(3x-5)}{3}$

$3 * (x + 2) < 2 * (3x - 5)$

$3x + 6 < 6x - 10$

Step 2: subtract 6 in both sides

$3x + 6 - 6 < 6x - 10 - 6$

$3x < 6x - 16$

Step 3: subtract $6x$ in both sides.

$3x - 6x < 6x - 6x - 16$

$-3x < -16$

Step 4: multiply with minus 1 in both sides.

The sign of the inequality will flip from < to >

$3x > 16$

Step 5: divide with the coefficient of the unknown.

$\frac{3x}{3} > \frac{16}{3}$

$x > 5\frac{1}{3}$

EXAMPLE

$\frac{1}{3}x + 3 > 9$

$\frac{1}{3}x + 3 - 3 > 9 - 3$

$\frac{1}{3}x > 6$

Multiply with 3 to cancel 1/3

$\frac{3*1}{3}x > 6 * 3$

$x > 9$

PRACTICE

1) Solve and check

$5x + 7 \geq 2$

2) Solve and graph the solution

$3x + 4 \geq 6 + 2x$

3) Solve

$\frac{2}{5}x - \frac{1}{2} > 3 + x$

4) Your school wants to raise money for charity. The school organizes a dance where the DJ costs $1200 and the ticket costs $8. How many tickets have to be sold to make a profit more than $1700?

a) Write an inequality to solve the problem

b) Solve and verify the solution

5) Solve

$3(x - 3) > \frac{2}{3}(3x + 6)$

6) John is replacing the light bulbs in his house from regular to energy saver light bulbs.

A regular light bulb costs $0.6 and has an electricity cost of $0.005 per hour.

An energy saver light bulb costs $5.5 and has an electricity cost of $0.001 per hour.

For how many hours of use it is cheaper to use an energy saver light bulb than a regular light bulb?

a) Write an inequality for this problem.

b) Solve the inequality. Explain the solution in words.

4.D. LINEAR INEQUALITIES WITH TWO VARIABLES

In this case we are working with x and y. At the same time the solutions are all the points that are situated in a certain area of a plane defined by the cartesian coordinates of the points.

EXAMPLE

Solve the inequality
$y < 2x + 1$

We have to find all the points in the plane that satisfy this condition.

Step 1:
Represent the straight line $y = 2x + 1$

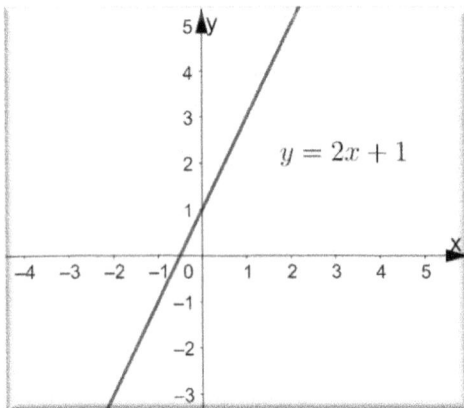

Step 2:
Check whether point (0,0) satisfies the inequality:
$y < 2x + 1$
We substitute zero for x and y respectively.
We have:
$0 < 2(0) + 1$
$0 < 1$
Because this is true, it means that point (0,0) is part of the solution.

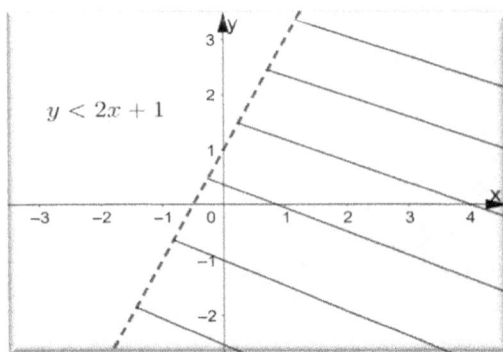

The fact that y is less than 2x+1 means that the points situated on the line
$y = 2x + 1$ are not part of the solution.
We represent the final solution as the area situated below the line $y = 2x + 1$
The line is a dashed line because the points on this line are not part of the solution. All the other points beneath the line $y = 2x + 1$ are part of the solution.

EXAMPLE

Solve the inequality

$y \geq -3x + 2$

We have to find all the points in the plane that satisfy this condition.

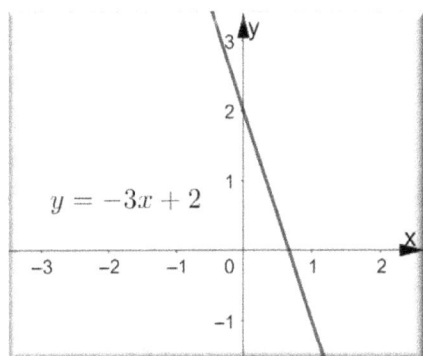

Step 1:

Represent the straight line $y = -3x + 2$

Step 2:

Check whether point (0,0) satisfies the inequality:

$y \geq -3x + 2$

We substitute zero for x and y respectively.

We have:

$0 \geq -3(0) + 2$

$0 \geq 2$

Because this is not true, it means that point (0,0) is not part of the solution.

The fact that y is greater and equal than -3x+2 means that the points situated on the line

$y = -3x + 2$ are part of the solution. The line will be a full line not dashed.

We represent the final solution as the area situated below the line $y \geq -3x + 2$

All the other points above the line $y = -3x + 2$ will be part of the solution.

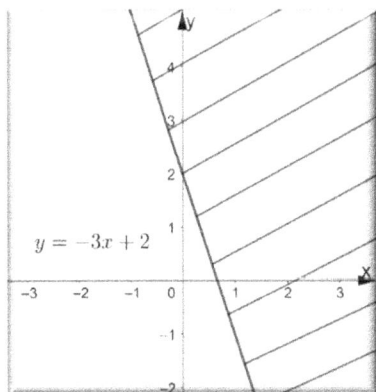

PRACTICE

1) Graph the solution

$y > 3x - 1$

2) Which point(s) is/are in the solution region of the inequality $3x - 6y \leq 4$

a) (0,0)　　　　b) (4,2)　　　　　　c) (-2,5)　　　　d) (3,-6)

3) In the graph below, the equation of the boundary line is: $x - 3y = 6$

Determine the inequality represented by the graph.

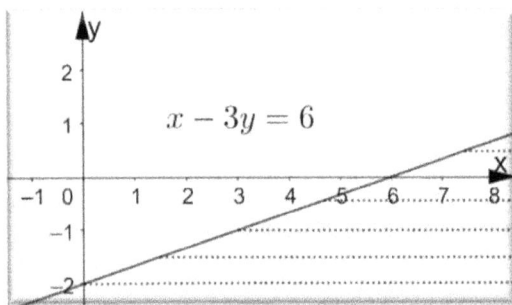

4) The graph below shows the solution to the inequality

a) Explain why the boundary line is a broken line?

b) Why the solution is beneath the line and not above?

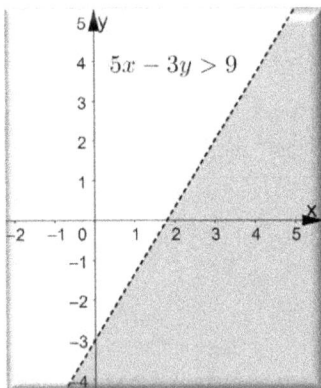

5) The point(s) which is NOT in the solution region of the inequality $5x - 2y > 5$ are:

a) (1,3)　　　　　　b) (3,2)　　　　　c) (2,-3)　　　　d(0,1)

Chapter 5

Polynomials

5.A. CHARACTERISTICS OF POLYNOMIALS

Definition of a polynomial.

An expression of more than two algebraic terms with different signs, especially the algebraic sum of several terms that contain different powers of the same variable(s)

Monomial is a polynomial with one term.

EXAMPLE

$3,\ 2x,\ 5x, 3x$, where; $3,\ 2x,\ 5x, 3x$ are **terms**

Binomial is a polynomial with two terms

EXAMPLE

$7 + 2x, 3x - 4, 5x + 6x^2, 2 - 4z$
7 *and* $2x$ are <u>terms</u> as well as all the others in the other binomials

Trinomial is a polynomial with three terms

EXAMPLE

$37 + 4x - 5x^2, 4x + 3x^2 + 2x^3, 6y^3 + 9y^2 - 77y, 27 - 45z - 3z^2$
Where any of $37, 4x, 5x^2$ are <u>terms</u> of the polynomial as well as all the others in the other trinomials.

Polynomial has more than three terms

EXAMPLE

$5x^3 + 3x^2 - 2x + 4$

Degree of the polynomial is the highest degree of the variables.
The degree of $5x^3 + 3x^2 - 2x + 4$ is three

A constant (a number) has the degree equal with zero.

The degree of $5x^3y^2 + 3x^2y - 2x + 4$ is five

5.B. ADDING AND SUBTRACTING POLYNOMIALS

a. Adding

Let's suppose we have to add the binomial $(2x - 1)$ with the binomial $(5x + 3)$

Like terms are the terms that have the same variable or group of variables at the same exponent, like: $2x + 4x$; $5x^2y + 4x^2y$; $7x^3u^2 - 6x^3u^2$

With like terms we add or subtract the coefficients, and then copy the variable or group of variables

Step 1 re-copy everything without the brackets
$(2x - 1) + (5x + 3) = 2x - 1 + 5x + 3$
Step 2 group the **like** terms
$2x - 1 + 5x + 3 = 2x + 5x - 1 + 3$
Step 3 add algebraically the like terms
$2x + 5x - 1 + 3 = 7x + 2$
So,
$(2x - 1) + (5x + 3) = 7x + 2$

EXAMPLE

$(4x + 6) + (-2x + 4) = 4x + 6 - 2x + 4 = 4x - 2x + 6 + 4 = 2x + 10$

b. Subtracting polynomials

Let's suppose we have to subtract the binomial $(3x - 4)$ from the binomial $(7x + 5)$
Step 1 re-copy the first polynomial because it has plus sign without the brackets, then write the second polynomial without the bracket but all the terms will have the opposite sign.
$(7x + 5) - (3x - 4) = 7x + 5 - 3x + 4$
Step 2 group the like terms
$7x + 5 - 3x + 4 = 7x - 3x + 5 + 4$
Step 3 add algebraically the like terms
$7x - 3x + 5 + 4 = 4x + 9$

So,

$(7x + 5) - (3x - 4) = 4x + 9$

EXAMPLE

1) $(2x + 4) - (-3x + 5) = 2x + 4 + 3x - 5 = 2x + 3x + 4 - 5 = 5x - 1$

2) $(3xy^2 + 4xy - 3x) - (4xy^2 - x^2y + 3xy + 7x - 4) = 3xy^2 + 4xy - 3x - 4xy^2 + x^2y - 3xy - 7x + 4 = 3xy^2 - 4xy^2 + 4xy - 3xy + x^2y - 3x - 7x + 4 = -4xy^2 + xy + x^2y - 10x + 4$

PRACTICE

1) Determine if the expressions below are a monomial, binomial or trinomial.

a) $3x - 2$, b) 4 c) $5x^2 + 3x - 1$ d) $x^2 - 3$

2) Determine the degree of the following polynomials

a) $5x^3 + 3x^2 - 2x + 4$ b) 5 c) $4x^2 - 5x + 6$

3) Add the polynomials:

a) $(2x^2 - 3x + 4) + (2x - 4) =$ b) $(3x^2 + 4x + 6) + (-4x - 5) =$

4) Subtract the polynomials

a) $(4x^2 - 3x + 2) - (3x^2 + 5x - 7) =$ b) $(2x^2 + 4x - 5) - (-3x^2 + 2x - 1) =$

5) Add and/or subtract the polynomials

a) $(4x^3 - 3x^2 - 2x + 1) + (2x^3 + 3x^2 - 4x - 5) =$

b) $(4x^3 - 2x + 1) - (2x^3 + 3x^2 - 4x) =$

6) Add and/or subtract the polynomials

$(4x^3 + 1) - (2x^3 + 3x^2) + (5x^2 - 6x - 4) =$

5.C. MULTIPLICATION OF POLYNOMIALS

Remember that a monomial is a polynomial with one term.

EXAMPLE

Multiply $2x \times 3x$
Step 1 Multiply the coefficients $2 \times 3 = 6$
Step 2 Multiply the variables $x \times x = x^2$
So,
$2x \times 3x = 6x^2$

EXAMPLE

Multiply $3xy^2 \times 4x^3y^3$
$3xy^2 \times 4x^3y^3 = 12x^{1+3}y^{2+3} = 12x^4y^5$

Expanding brackets means getting rid of them.
$a \times (b + c) = a \times b + a \times c$

EXAMPLE

$2 \times (7 + 4) = 2 \times 7 + 2 \times 4 = 14 + 8 = 22$

EXAMPLE

$6(2 + 5 - 2x) = 6 \times 2 + 6 \times 5 - 6 \times 2x = 12 + 30 - 12x = 42 - 12x$

FOIL (First, Outer, Inner, Last)

First, we multiply the first terms in each binomial. Then we multiply the Outer which means that we multiply the outermost terms in the product. Then Inner terms and Outer ones.

EXAMPLE

$(a + b)(c + d) = ac + ad + bc + bd$
We multiply the First terms $\boldsymbol{a} \times \boldsymbol{c}$
then we multiply the Outer terms $\boldsymbol{a} \times \boldsymbol{d}$,
We multiply the Inner terms $\boldsymbol{b} \times \boldsymbol{c}$,

and then we multiply the <u>Last</u> terms $b \times d$

We apply here the same strategy.

Multiply:

$(2x - 3) \times (4x + 5) = 2x \times 4x + 2x \times 5 - 3 \times 4x - 3 \times 5 = 8x^2 + 10x - 12x - 15 = 8x^2 - 2x - 15$

EXAMPLE

Multiply:

$-(7x + 6) \times (5x - 4) = [-(7x \times 5x - 7x \times 4 + 6 \times 5x + 6 \times (-4)] = -(35x^2 - 28x + 30x - 24) = -(35x^2 + 2x - 24) = -35x^2 - 2x + 24$

When we have to multiply a polynomial with another polynomial, we follow the following steps:

Step 1

Multiply first term of the first polynomial with all the terms of the second.

Step 2

Multiply second term of the first polynomial with all the terms of the second.

Step 3

Multiply third term of the first polynomial with all the terms of the second.

Step 4

Multiply all the other terms of the first polynomial with all the terms of the second like above.

Step 4

Add or subtract like terms

EXAMPLE

Multiply: $(2x^2 - 3x + 4)(-3x^2 + 4x - 5)$

$(2x^2 - 3x + 4)(-3x^2 + 4x - 5) = 2x^2 \times (-3x^2) + 2x^2 \times 4x + 2x^2 \times (-5) + (-3x) \times (-3x^2) + (-3x) \times (4x) + (-3x) \times (-5) + 4 \times (-3x^2) + 4 \times 4x + 4 \times (-5) = -6x^4 + 8x^3 - 10x^2 + 9x^3 - 12x^2 + 15x - 12x^2 + 16x - 20 = -6x^4 + 17x^3 - 34x^2 + 31x - 20$

PRACTICE

1) $(3x^2 - 2)(x + 1) =$

2) $(x^2 + 3)(4x - 5) =$

3) $(xy - x^2 + 3)(y + 3xy - 2) =$

4) $(x - 1)(x + 1)(2xy + x + y) =$

5) $(2x - 3)(2x + 3)(x + x^2 - x^3 + 4) =$

5.D. RATIONALIZING THE DENOMINATOR, SPECIAL BINOMIAL PRODUCTS

Remember that when we multiply the same radical by itself as many times as the root order, the result is the number under the radical.

EXAMPLE

$\sqrt{3} \times \sqrt{3} = 3$

$\sqrt[3]{5} \times \sqrt[3]{5} \times \sqrt[3]{5} = 5$

Rationalizing the denominator

a. When we have a fraction that has a radical at the denominator, we can transform the radical into a non-radical by multiplying the same radical by itself as many times as the root order, the result is the number under the radical.

EXAMPLE

Rationalize the denominator.

$\frac{5}{\sqrt{2}} = \frac{5 \times \sqrt{2}}{\sqrt{2} \times \sqrt{2}} = \frac{5\sqrt{2}}{2}$

$\frac{2x}{\sqrt{x+1}} = \frac{2x \times \sqrt{x+1}}{\sqrt{x+1} \times \sqrt{x+1}} = \frac{2x\sqrt{x+1}}{x+1}, x+1 > 0$

b. Using the special binomial product; the difference of squares

$(a + b) \times (a - b) = a^2 - ab + ab - b^2 = a^2 - b^2$

So,

$(a + b) \times (a - b) = a^2 - b^2$

EXAMPLE

Rationalize the denominator.

$\frac{\sqrt{3}}{\sqrt{5}-\sqrt{3}} = \frac{\sqrt{3} \times (\sqrt{5}+\sqrt{3})}{(\sqrt{5}-\sqrt{3}) \times (\sqrt{5}+\sqrt{3})} = \frac{\sqrt{15}+\sqrt{3} \times \sqrt{3}}{[(\sqrt{5})^2 - (\sqrt{3})^2]} = \frac{\sqrt{15}+3}{5-3} = \frac{\sqrt{15}+3}{2}$

Other two special binomial products

$(a + b)^2 = (a + b) \times (a + b) = a^2 + 2ab + b^2$

$(a - b)^2 = (a - b) \times (a - b) = a^2 - 2ab + b^2$

EXAMPLE

$(x + 2)^2 = (x + 2) \times (x + 2) = x^2 + 2x \times 2 + 2^2 = x^2 + 4x + 4$

$(3x - 5)^2 = (3x - 5) \times (3x - 5) = 9x^2 - 2 \times 3x \times 5 + 5^2 = 9x^2 - 30x + 25$

EXAMPLE

Rationalize the denominator.

$\frac{1}{\sqrt{x^2+4x+4}} = \frac{1}{\sqrt{(x+2)^2}} = \frac{1}{x+2} \quad x \neq -2$

PRACTICE

Determine the answer.

1) The rationalized expression of $\frac{2}{\sqrt{7}}$ is

2) The rationalized expression of $\frac{2\sqrt{3}}{\sqrt{11}}$ is

3) The rationalized expression of $\frac{\sqrt{7}+3\sqrt{5}}{\sqrt{3}}$ is

4) The rationalized expression of $\frac{\sqrt{5}}{\sqrt{3}+\sqrt{2}}$ is

5) The area of a rectangle is $\sqrt{7} + 2\sqrt{5}$ and length $\sqrt{3} - 1$. The width is:

5.E. APPLICATIONS OF POLYNOMIALS

a. Area

Sometimes, the dimensions of the geometric shapes have variables in their relation

EXAMPLE

Find the area for the rectangle below.

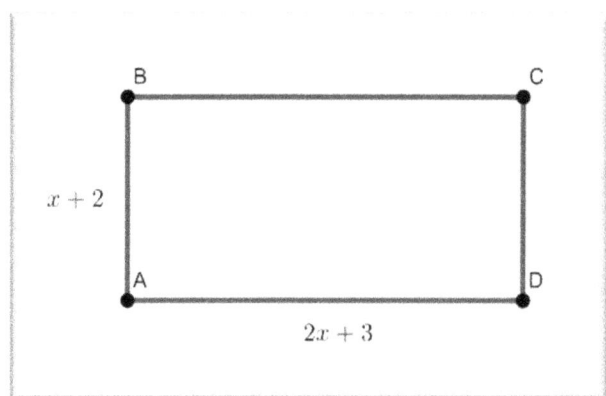

$Area = Base \times heigh = AD \times AB =$

$(2x + 3)(x + 2) = 2x \times x + 2x \times 2 + 3 \times x +$

$3 \times 2 = 2x^2 + 4x + 3x + 6 = 2x^2 + 7x + 6$

b. Perimeter

The perimeter equals the sum of all the sides of the geometric shape
In the rectangle above the perimeter is calculated:
$P = 2x + 3 + x + 2 + 2x + 3 + x + 2 = 6x + 10$

c. Volume

For example, the volume of the prism below is:
$Volume = Area\ base \times height = (x + 1)(x - 5)(x) = (x^2 - 5x + x - 5)(x) = (x^2 - 4x - 5)(x)$
$$= x^3 - 4x^2 - 5x$$

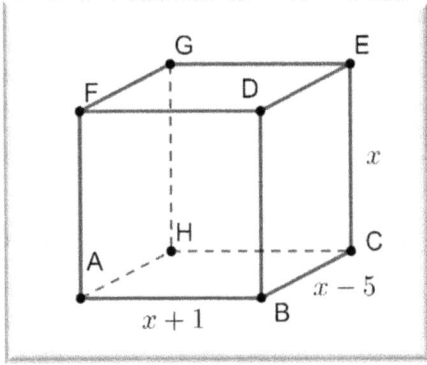

d. Profit = revenue − cost

The cost for producing an object is \$2. The profit has a linear increase $y = 3x + 2$

Calculate the profit at 3 days.

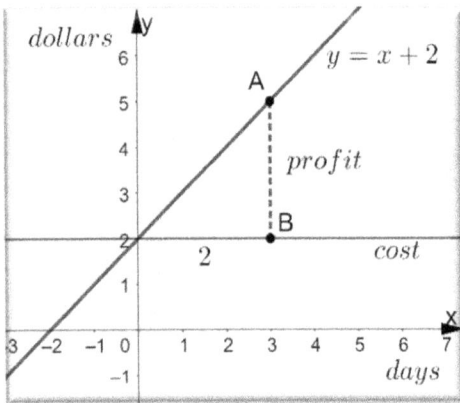

Profit=Revenue - Cost

$Revenue(3\ days) = x + 2 = 3 + 2 = \$5$

Cost =\$2

$Profit = \$5 - \$2 = \$3$

PRACTICE

1) Calculate the area of the figure below

EC=2

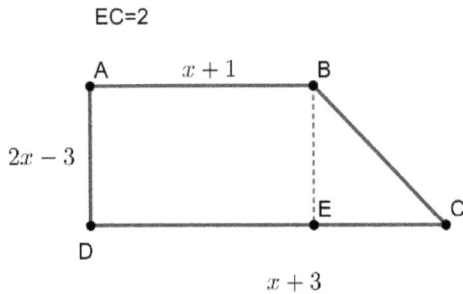

2) Calculate the area of the figure below

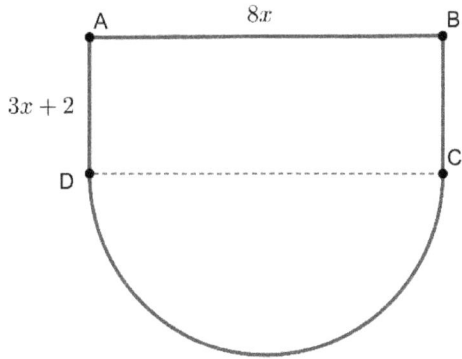

$8x$

A   B

$3x + 2$

D   C

3) Calculate the volume of the figure below

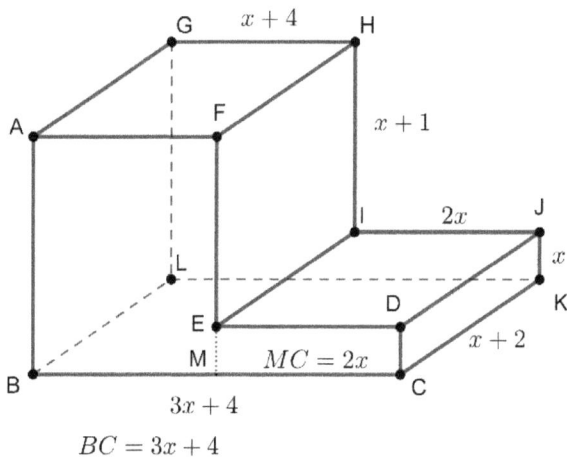

$x + 4$

G   H

$x + 1$

A   F

$2x$

I   J

x

L   D   K

E

M $MC = 2x$ $x + 2$

B   C

$3x + 4$

$BC = 3x + 4$

4) Calculate the perimeter of the figure below

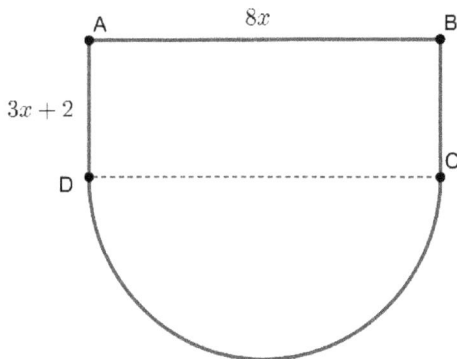

$8x$

A   B

$3x + 2$

D   C

5) Calculate the perimeter of the figure below, BC = 9

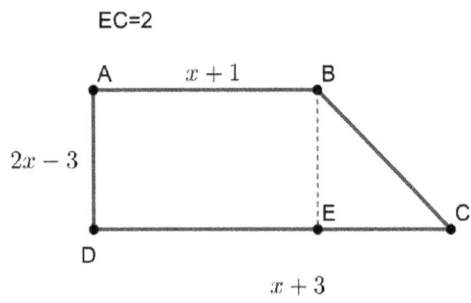

EC=2

A $\quad x+1 \quad$ B

2x − 3

E $\qquad$ C

D

x + 3

5.F. DIVISION OF POLYNOMIALS

Remember when we divide say, 346 by 15 we have:

$$
\begin{array}{r}
2\quad 3 \\
15\ \overline{\big)\ 3\quad 4\quad 6} \\
-\ 3\quad 0 \\
\overline{\quad 4\quad 6} \\
-\ 4\quad 5 \\
\overline{\quad 1}
\end{array}
$$

We can write this division as:

$$\frac{Dividend}{Divisor} = Quotient + \frac{Remainder}{Divisor}$$

Here we have:

$$\frac{346}{15} = 23 + \frac{1}{15}$$

If we have again

$\frac{Dividend}{Divisor} = Quotient + \frac{Remainder}{Divisor}$ and multiply both sides with the Divisor

We have:

$Dividend = Quotient * Divisor + Remainder$

Let us divide now two polynomials.

$3x^2 + 2x - 4 \ divided \ by \ x + 3$

Repeat the steps until the degree of the remainder is less than the degree of the divisor.

$$
\begin{array}{r}
3x\ -7 \\
x+3\ \overline{\big)\ 3x^2 + 2x - 4} \\
-3x^2 - 9x \\
\overline{=\quad -7x - 4} \\
7x + 21 \\
\overline{17}
\end{array}
$$

$Dividend = Quotient * Divisor + Remainder$

$3x^2 + 2x - 4 = (3x - 7) * (x + 3) + (17)$

a. Division by a quadratic polynomial

Divide:

$4x^4 - x^3 + 2x^2 - 3x + 5 \div x^2 + 1$

$$
\begin{array}{r}
4x^2 - x + 6 \\
x^2 + 1 \overline{\smash)\; 4x^4 - x^3 + 2x^2 - 3x + 5} \\
-4x^4 - 0 + 4x^2 \\
\hline
-x^3 + 6x^2 - 3x \\
x^3 \;\; + 0 \;\; + x \\
\hline
6x^2 - 2x + 5 \\
-6x^2 \;\; - 0 \;\; - 6 \\
\hline
-2x - 1
\end{array}
$$

PRACTICE

1) Divide: $x^2 - 3x + 4 \div x - 2$

2) Divide $2x^3 - 3x^2 + 4x - 5 \div x + 4$

3) Divide $5x^4 - 4x^3 + 3x^2 - 2x + 1 \div x^2 + 2x - 3$

4) Divide $4x^4 + 2x^2 - x + 5 \div 2x^2 - x + 4$

5) Divide $x^3 - 3 \div x + 2$

b. Synthetic Division

This is a different method used instead of long division, using only the coefficients of the terms of the dividend, <u>when the divisor is a binomial of the form (x−n)</u>.

EXAMPLE

Divide $3a^2 - 4a + 5$ by $a - 2$

We will always use the coefficients of the dividend; 3,-4,5, and the opposite number of n.

```
2 | 3  -4   5
  |     6   4
  |_____
    3   2   9
```

$$3a^2 - 4a + 5 \div a - 2 = 3a + 2 + \frac{9}{a-2}$$

PRACTICE

1) Divide $4x^3 - 3x^2 + 2x + 1 \div x - 3$

2) Determine the quotient and remainder for $x^3 - 4x + 3 \div x + 2$

3) Divide $2x^4 + 3x^2 - 4x + 1 \div x + 1$

4) Divide $x^4 - 2x^3 - x^2 + 3x - 4 \div x - 2$

5. Divide $x^2 - 1 \div x + 5$

CHAPTER 6

Factoring Polynomials

6.A. COMMON FACTORS OF POLYNOMIALS

Let us suppose we have the product $6 = 2 \times 3$.
Here, 2 and 3 are called factors of the multiplication

The common factor that 6 and 9 have is 3. Why?
$6 = 2 \times \mathbf{3}$
$9 = 3 \times \mathbf{3}$

We see that these two numbers, 6 and 9, have factor 3 common.

Let us have a polynomial
$2x^2 + 4x - 6$
Here, 2 is the common factor of all the terms of the polynomial.
We can write it as:
$2(x^2 + 2x - 3)$

6.B. FACTORING POLYNOMIALS OF THE FORM $x^2 + bx + c$

One of the methods used to factor polynomials of the form $x^2 + bx + c$ is by noticing that
    a) two numbers when multiplied give us the constant "c"
    b) the same two numbers when added give us the coefficient "b"
Those numbers, say m and n, are the numbers in the factoring formula:
$(x + m)(x + n)$

EXAMPLE

Factor $x^2 + 3x + 2$
The numbers that multiplied would give us 2 are 1 and 2 respectively.
If we add 1 and 2 we get 3, the coefficient of x.
So,
$x^2 + 3x + 2 = (x + 2)(x + 1)$

NOTE

There is important to know what we should look for when we have different signs for coefficient "b" and constant "c"

PRODUCT or CONSTANT "c"	The sign of the integer numbers
POSITIVE	Both numbers negative
	Both numbers positive
NEGATIVE	One number negative
	The other positive

PRACTICE

1) Factor $x^2 + 6x + 8$

2) Factor $x^2 - 9x + 20$

3) Factor $x^2 - 5x - 24$

4) Factor $x^2 + 10x + 21$

5) Factor $x^2 - 10x + 9$

6.C. FACTORING POLYNOMIALS $ax^2 + bx + c$ BY DECOMPOSITION

Let us suppose that we want to factor the polynomial of the form $ax^2 + bx + c$

EXAMPLE

$10x^2 + 7x - 12$

We have to find:
  a) two numbers that multiplied give us "**a×c**"
  b) the same two numbers added give us the coefficient "**b**"

Here:

$10 \times (-12) = -120$

$120 = 2 \times 6 \times 10 = 2 \times 2 \times 3 \times 2 \times 5$

Let's try $8 \times 15 = 120$

Remember that one integer is negative and the other positive.

So,

The sum will give us 7. One is -15, the other is 8

We will **DECOMPOSE** $10x^2 + 7x - 12$ with:

$10x^2 - 15x + 8x - 12 = 5x(2x - 3) + 4(2x - 3) = (2x - 3)(5x - 4)$

PRACTICE

1) Factor $5x^2 + 2x - 3$

2) Factor $3x^2 - 2x - 8$

3) Factor $15x^2 - 22x + 8$

4) Factor $12x^2 + 28x - 5$

5) Factor $21x^2 - 8x - 45$

6.D. FACTORING POLYNOMIALS IN DIFFERENCE OF SQUARE FORM

When the polynomial is a binomial with the form of $(ax)^2 - b^2$; *a and b integers*:
We can write this as $(ax)^2 - b^2 = (ax - b)(ax + b)$.

EXAMPLE

Factor $4x^2 - 1$
Here, $4x^2 = 2^2x^2 = (2x)^2$

$4x^2 - 1 = (2x)^2 - 1^2 = (2x - 1)(2x + 1)$

PRACTICE

1) Factor $x^2 - 16$

2) Factor $25x^2 - 9$

3) Factor $36x^2 - 49$

4) Factor $2x^2 - 100$

5) Factor $9x^2 - 5$

6.E. SOLVING A QUADRATIC EQUATION BY FACTORING

IMPORTANT

If $c * b = 0$, then we know that it is either c=0 or b=0. (This includes the possibility that they are both 0.)

That is; if <u>two factors are multiplied and have the result 0, **at least** one of the factors must be equal to 0.</u>

EXAMPLE

Solve:

$(x - 4) * (2x + 6) = 0$

This product is zero if

$x - 4 = 0$ Results $x = 4$

Or,

$2x + 6 = 0$ Results, $2x = -6 \Rightarrow x = \frac{-6}{2} = -3$

It is possible to solve a quadratic equation if one side of the equation is factored. For this to be effective, the other side **must equal** 0.

A **solution** of an equation is a value that satisfies the equation.

That is, a solution is a value that, if substituted into the equation, will make the equation true, i.e. the left side of the equation will be equal with the right side of the equation.

A quadratic relation can be written as $y = ax^2 + bx + c$ where a, b, and c are real numbers and a≠0.

Here are some examples of quadratic equations in x:

- $x^2 + 2x = 1$

- $(3x - 4)(2x + 5) = 0$
- $-4(x^2 + 1) + 7 = 2$
- $x^2 = 9x - 5$

A quadratic equation can have 0, 1, or 2 real roots (i.e., solutions that are real numbers).

Sometimes quadratic equations have imaginary solutions, but we won't worry about that for now.

EXAMPLE

Solve:

$x^2 + 3x - 10 = 0$

Factoring the left side, we have

$(x - 2)(x + 5) = 0$

So,

$x - 2 = 0$

Or,

$x + 5 = 0$

Graphing Factored Form When the Zeros Are Integers

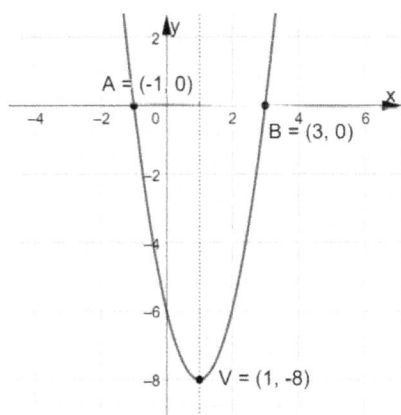

Graph $y = 2(x + 1)(x - 3)$

$x_{symetry} = \dfrac{-1+3}{2} = \dfrac{+2}{2} = 1$

$y_{vertex} = 2(1 + 1)(1 - 3) = 2(2)(-2) = -8$

Graphing Factored Form When the Zeros Are Not Integers

Graph

$$y = -\frac{1}{5}(x - 5)(2x + 4)$$

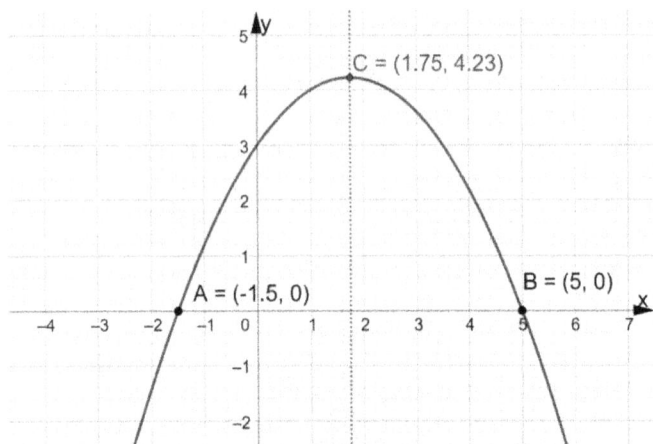

PRACTICE

1) Solve $x^2 + 6x + 8 = 0$

2) Solve: $x^2 - 36 = 0$

3) Solve $(x - 1)^2 = 0$

4) Solve $3x^2 + 15x + 12 = 0$

5) Solve

$6x^2 + 13x - 5 = 0$

6.F. QUADRATIC FORMULA

a. Solve by completing the square

Let us suppose we start from the general form

$$ax^2 + bx + c = 0, a \neq 0$$

The idea is to use first two terms $ax^2 + bx$ and transform them into the form

$x^2 + 2kx + k^2$ that will become $(x + k)^2$, where k is a real number

Divide by a

$$x^2 + \frac{b}{a}x + \frac{c}{a} = 0$$

We multiply and divide by 2 the term with x

$$x^2 + 2 * \frac{b}{2a}x + \frac{c}{a} = 0$$

We add and subtract $\frac{b^2}{4a^2}$

$$x^2 + 2 * \frac{b}{2a}x + \frac{b^2}{4a^2} - \frac{b^2}{4a^2} + \frac{c}{a} = 0$$

We group the first three terms as $(x + \frac{b}{2a})^2$, where $k = \frac{b}{2a}$

$$(x + \frac{b}{2a})^2 - \frac{b^2}{4a^2} + \frac{c}{a} = 0$$

We compute $-\frac{b^2}{4a^2} + \frac{c}{a} = -\frac{b^2}{4a^2} + \frac{4ca}{4a^2} = \frac{-b^2 + 4ac}{4a^2}$

$$(x + \frac{b}{2a})^2 + \frac{-b^2 + 4ac}{4a^2} = 0$$

Subtract $\frac{-b^2 + 4ac}{4a^2}$ on both sides.

$$(x + \frac{b}{2a})^2 = \frac{b^2 - 4ac}{4a^2}$$

We take the square root on both sides.

$$\sqrt{(x + \frac{b}{2a})^2} = \pm\sqrt{\frac{b^2 - 4ac}{4a^2}}$$

$$x + \frac{b}{2a} = \pm\sqrt{\frac{b^2 - 4ac}{4a^2}}$$

We subtract $\frac{b}{2a}$ on both sides

$$x = -\frac{b}{2a} \pm \sqrt{\frac{b^2 - 4ac}{4a^2}}$$

$$x = -\frac{b}{2a} \pm \frac{\sqrt{b^2 - 4ac}}{2a}$$

Quadratic Formula

$$x = \frac{-b \pm \sqrt{b^2 - 4ac}}{2a}$$

EXAMPLE

Solve by completing the square:

$3x^2 + 2x - 4 = 0$

Divide by 3

$x^2 + 2 * \frac{1}{3}x - \frac{4}{3} = 0$

Add and subtract $\frac{1}{9}$

$x^2 + 2 * \frac{1}{3}x + \frac{1}{9} - \frac{1}{9} - \frac{4}{3} = 0$

Compute $-\frac{1}{9} - \frac{4}{3} = -\frac{1}{9} - \frac{12}{9} = \frac{-1-12}{9} = \frac{-13}{9}$

$x^2 + 2 * \frac{1}{3}x + \frac{1}{9} - \frac{13}{9} = 0$

Add $\frac{13}{9}$ on both sides.

$x^2 + 2 * \frac{1}{3}x + \frac{1}{9} = \frac{13}{9}$

$(x + \frac{1}{3})^2 = \frac{13}{9}$

We take the square root on both sides.

$\sqrt{(x + \frac{1}{3})^2} = \pm\sqrt{\frac{13}{9}}$

$x + \frac{1}{3} = \pm\sqrt{\frac{13}{9}}$

We subtract $\frac{1}{3}$ on both sides

$$x = -\frac{1}{3} \pm \sqrt{\frac{13}{9}}$$

PRACTICE

1) The number to be added to $x^2 + 5x$ to make a perfect square is:

2) The term outside the square of expression $x^2 - 5x + a$ is:

3) The completed square form of $x^2 + 6x - 4$ is

4) The completed square form of $ax^2 + 2x - 4, a \neq 0$ is

5) To form a perfect square k in $x^2 + kx + 3$, $k =$

b. Roots of Quadratic Equations

We can use two methods

a) Factoring

b) Using Quadratic Formula

What does it mean to find the roots or the zeros of the equation $ax^2 + bx + c = 0, a \neq 0$?

It means that we are finding the values of x, or the x coordinates of the intersection of the parabola with the x axis.

Remember that any point on the x axis has the coordinates (x,0) Equalizing the quadratic polynomial with zero, we are finding the values of x for which the y is zero.

EXAMPLE

Find the roots:

$$x^2 - 2x - 24 = 0$$

Using factoring, we have:

$$x^2 - 2x - 24 = (x - 6)(x + 4) = 0$$

$$x - 6 = 0 \Rightarrow x = 6$$

Or,

$$x + 4 = 0 \Rightarrow x = -4$$

What we found are the points of intersection between the parabola $y = x^2 - 2x - 24$ and x axis or

$(-4,0)$ and $(6,0)$

Find the roots:

$$x^2 - 4x - 7 = 0$$

We can't factor so, we will use the quadratic formula

$$x = \frac{-b \pm \sqrt{b^2 - 4ac}}{2a}$$

Here, a=1; b=-4; c=-7

$$x = \frac{-b \pm \sqrt{b^2 - 4ac}}{2a} = \frac{-(-4) \pm \sqrt{(-4)^2 - 4(1)(-7)}}{2} = \frac{4 \pm \sqrt{16 + 28}}{2} = \frac{4 \pm \sqrt{44}}{2} = 2 \pm \frac{6.63}{2} = 2 \pm 3.31$$

c. The Discriminant $b^2 - 4ac$

1. $b^2 - 4ac > 0$ the equation will have **two** solutions

2. $b^2 - 4ac = 0$ the equation will have **one** solution

3. $b^2 - 4ac < 0$ the equation will have **NO** REAL solution

PRACTICE

1) Find the number of real solutions the equation $x^2 - 4x + 7 = 0$ has.

2) Find the number of real solutions the equation $2x^2 - 3x + 4 = 0$ has.

3) Find the number of real solutions the equation $3x^2 + 4x - 1 = 0$ has.

4) Find the number of real solutions the equation $x^2 - 4x + 4 = 0$ has.

5) Find the number of real solutions the equation $5x^2 + 4x - 9 = 0$ has.

CHAPTER 7

Functions

7.A. PARENT FUNCTIONS

a. Determine: if a relation is a function, the values of a function, the range

A relation is a <u>function</u> when for each value x that belongs to the domain, there is <u>only one value</u> y that belongs to the range.

EXAMPLE

Through function f in the picture below, there is only one value in the Range that corresponds to any value in the Domain. For value 2 in the Domain, there is only one value in the Range (4). For value 4 in the Domain, there is only one value in the Range (4). For value 3 in the Domain, there is only one value in the Range (9).

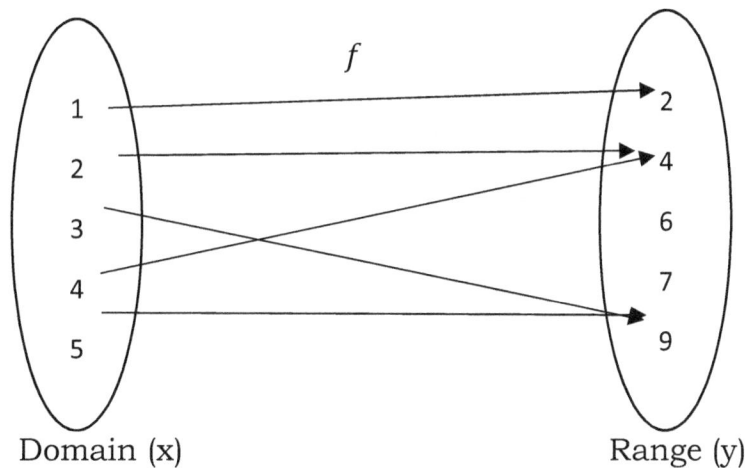

Domain (x)   Range (y)

The value of a function is the value in the Range that is the result of a value in the Domain going through the function f.

EXAMPLE

In the figure above one value of the function f is 9. This number is the result of the domain value of 3 going through f as well as 5 going through f.
The <u>Range</u> is the set of elements (letters or numbers) that are the result of the elements (letters or numbers) in the Domain going through function f.

EXAMPLE

The set of numbers {2,4,6,7,9} in the figure above is the Range of f.
The Domain is the set of numbers {1,2,3,4,5}.

PRACTICE

Determine if the relations below are functions.

1) {(-3, 6), (-2, 10), (3, 3), (3, -12), (7,12)}

{(-4, 8), (-4, 1), (-2, 3), (0, -12), (1,2),(2, 3)}

2)

X	Y
5	5
6	7
7	8
9	10

3) Determine if the values for each of the following functions are correct.

$f(x) = 5x + 3$ $\qquad$ $f(2)=15$

$f(x) = 3x^2 - 3x + 2$ $\qquad$ $f(1) = 2$

$G(s) = \dfrac{s^2-3s+7}{s+3}$ $\qquad$ $G(2) = 1$

4) Determine if the range of the following functions is correct. The domain is given.

$G(t) = 3 - 2t$ $\qquad$ D = {-1, -2, 3} $\qquad$ R= {5, 7, -3

$F(x) = x^2 - 5x + 1$ $\quad$ D = {-2, 0, 3} $\qquad$ R= {15, 1, -3}

$H(c) = \dfrac{c^2-2c}{c+2}$ $\qquad$ D = {-3, 0, 3} $\qquad$ R= {-15, 0, $\dfrac{3}{5}$ }

b. Linear and quadratic functions and their graphs

A function is <u>linear</u> when the difference between consecutive values in the Domain is always the same. At the same time, the difference between consecutive values in the Range is always the same, not necessarily the same with the difference between consecutive Domain values.

EXAMPLE

Domain(X)	Range(Y)	Point
1	5	A
2	7	B
3	9	C
4	11	D

Here the difference between consecutive values in the Domain is 1. $x_B - x_A = 2 - 1 = 1$, or $x_D - x_C = 4 - 3 = 1$.

At the same time the difference between consecutive values in the Range is 2. $y_B - y_A = 7 - 5 = 2$, or $y_D - y_C = 11 - 9 = 2$.

When we graph a linear function, the graph is a straight line.

The pairs (x,y) represent points in the Cartesian system of axes.

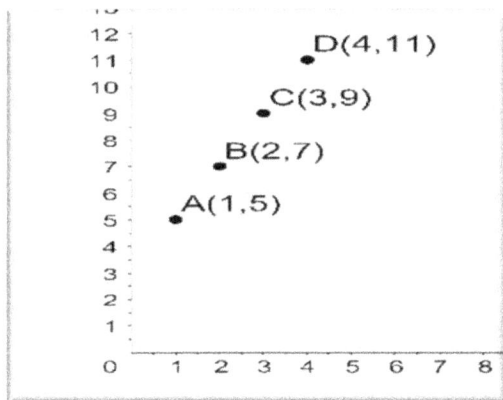

We can see here that if we connect these points we will create a straight line.

The equation that represents this graph is $y = 2x + 3$ and it is called a linear equation (the degree of the equation is either 0 or 1).

A function is *quadratic* when the relation between x and y is a polynomial of second degree.

$$y = x^2 - 5x + 4 \quad \text{or} \quad xy + 2x^2 - 3y = 5$$

EXAMPLE

$$y = x^2 - 5x + 4$$

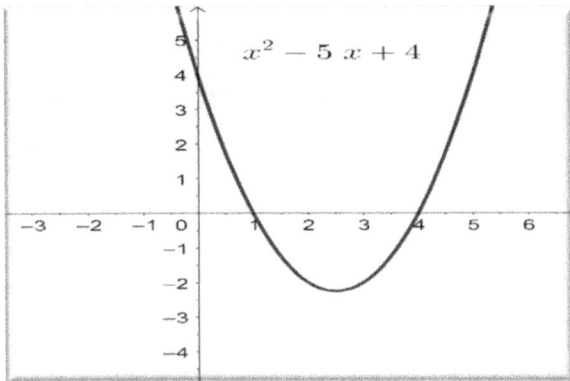

Here the quadratic function $y = x^2 - 5x + 4$ is represented.

VERTICAL LINE TEST

If we draw a vertical line through a graph and this line **intersects the graph** in **only one point**, then the graph represents a **function**. If this line **intersects the graph** in **more than one point**, then the graph DOES NOT represent a Function.

PRACTICE

Determine if the following tables of pairs x, y and expressions represent a linear function.

1)

X	Y
2	0
2	5
3	10
4	15

2)

X	Y
5	5
6	7
7	8
9	10

Determine if the following graph represents a linear function.

3)

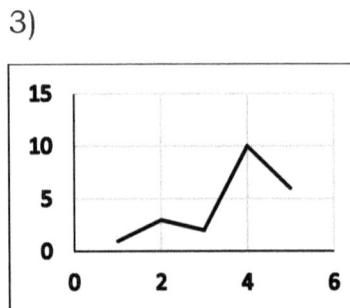

Determine if the following expressions represent a quadratic function.

4) $f(x) = 2x^2 - 3x + 4$ $f(x) = x^2 - 3x$ $f(x) = x^4 - 3x + 4$

c. Inverse functions and their graphs

An inverse function represented by $f^{-1}(x)$ is the function that has the Domain equal with the Range of the original function $f(x)$. The Range of the inverse function equals the Domain of the original function.

Domain and Range of inverses of functions

An inverse function represented by $f^{-1}(x)$ is the function that has the Domain equal with the Range of the original function $f(x)$. The Range of the inverse function equals the Domain of the original function.

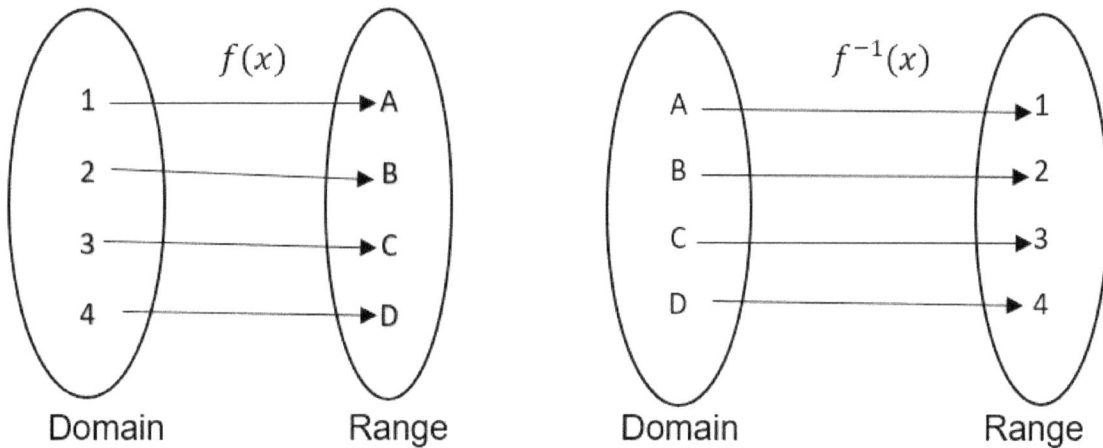

The Domain of $f(x)$ is: $D_{f(x)} = \{1,2,3,4\}$

The Range of $f(x)$ is: $R_{f(x)} = \{A, B, C, D\}$

The Domain of $f^{-1}(x)$ is: $D_{f^{-1}(x)} = \{A, B, C, D\}$

The Range of $f(x)$ is: $R_{f^{-1}(x)} = \{1,2,3,4\}$

EXAMPLE

A = {(-3, 5), (-2, 10), (3, 3), (3, -12), (7,12)}

B = {(-3, 6), (-2, 10), (0, 3), (3, -12), (7,22)}

For the set B above: $\{-3, -2, 0, 3, 7\}$ is the DOMAIN

{3,6,10,12,22} is the RANGE

EXAMPLE

We obtain the equation of the inverse function in a few steps:

1. Write the equation of the original function.

The original function is $f(x) = y = 4x + 7$

2. Switch the variables x and y in the original formula.

$x = 4y + 7$

3. Solve for y

$x - 7 = 4y$ so, $y = f^{-1}(x) = \frac{x-7}{4}$

EXAMPLE

State the Domain and the Range of the following

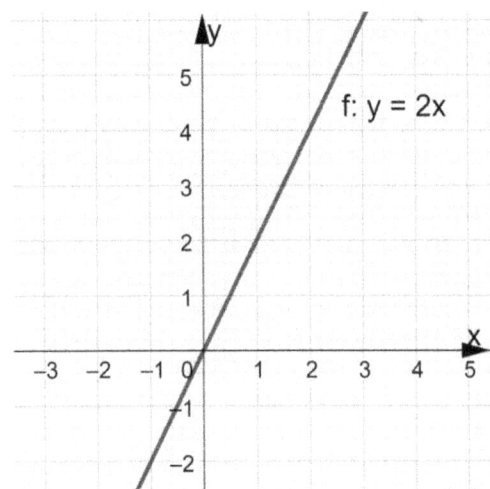

f: y = 2x

Domain is: $\{x | x \in R\}$

The domain is the set of values x <u>such that</u> x is an element of the set of real numbers.

Range is: $\{y | y \in R\}$

The range is the set of values y <u>such that</u> y is an element of the set of real numbers.

EXAMPLE

$f(x) = 2x + 2$

$f^{-1}(x) = \frac{1}{2}x - 1$

	$f(x) = 2x + 2$	$f^{-1}(x) = \frac{1}{2}x - 1$
Domain	$-3 \leq x \leq 2$	$-4 \leq x \leq 5$
Range	$-4 \leq x \leq 5$	$-3 \leq x \leq 2$

d. Inverses of linear and quadratic functions

The inverse of a function is the relation which undoes the work of the function.

EXAMPLE

The inverse of $f(x) = 2x - 7$ is written as $f^{-1}(x)$. Find it.

$y = 2x - 3$

Step 1 we interchange the x with y

$x = 2y - 3$

Step 2 we isolate y

$x + 3 = 2y - 3 + 3$ we add 3 on both sides

$x + 3 = 2y$

Or,

$2y = x + 3$ we divide by 2 on both sides

$y = \frac{1}{2}x + \frac{3}{2}$

So,

$f^{-1}(x) = \frac{1}{2}x + \frac{3}{2}$

$f(x) = y = 2x - 7$

We can graph both the function and the inverse on the same graph.

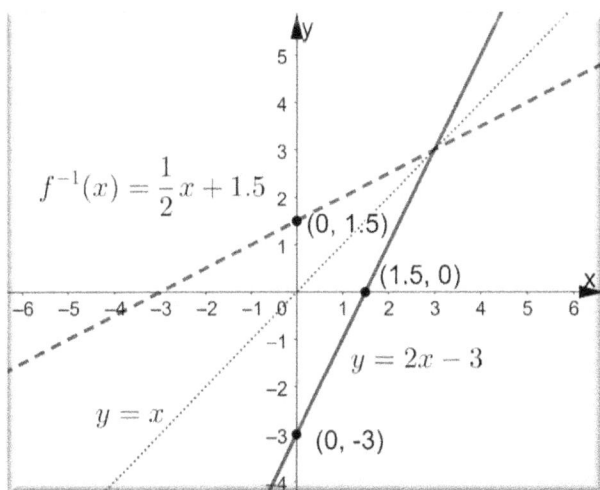

The inverse of the vertical line x=a is the horizontal line y=a.

The inverse of the horizontal line y=a is the vertical line x=a. In this case, the inverse is not a function.

The graph of an inverse function is always a reflection of the graph of the original function by y=x.

EXAMPLE

$f(x) = 4x + 1 \text{ and } f^{-1}(x) = \frac{x-1}{4}$

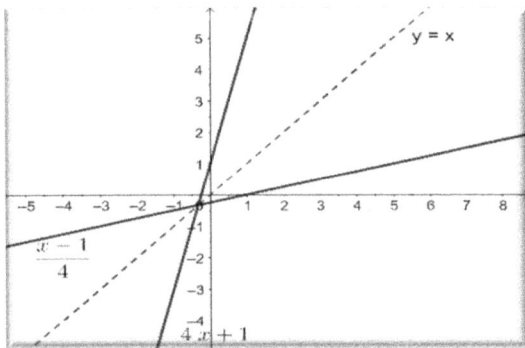

PROPERTY OF INVERSE FUNCTIONS

$$f\left(f^{-1}(x)\right) = f^{-1}(f(x)) = x$$

The function of the inverse function for a value x, equals the inverse function of the same function for value x, and equals value x.

PRACTICE

Determine the inverse function.

1) If $f(x) = y = 3x - 5$ the inverse function is: $f^{-1}(x) =$

2) If $f(x) = y = \frac{3}{2x+4}$ the inverse function is: $f^{-1}(x) = y =$

3) If $f(x) = y = \frac{\sqrt{x-1}}{3}$ the inverse function is: $f^{-1}(x) = y =$

4) If $f(x) = y = \frac{3x}{x+3}$ the inverse function is: $f^{-1}(x) =$

5) If $f(x) = y = \frac{1}{5x+3}$ the inverse function is: $f^{-1}(x) =$

7.B. POLYNOMIAL FUNCTIONS

Our focus for this unit will be polynomial functions and the related polynomial equations.

A **polynomial function** is a function whose equation is defined by a polynomial in **one variable**. The general form of a polynomial function is

$$f(x) = c_n x^n + c_{n-1} x^{n-1} + c_{n-2} x^{n-2} + \cdots + c_1 x + c_0$$

Where the coefficients $c_0, c_1, \ldots \ldots c_n$ are **real numbers** and the exponents of x are whole numbers (non-negative integers)

Terminology

The numerical coefficient of the highest degree term in a polynomial is called the leading coefficient.

In the general function,

$$f(x) = c_n x^n + c_{n-1} x^{n-1} + c_{n-2} x^{n-2} + \cdots + c_1 x + c_0$$

The coefficient c_n is the leading coefficient.

The domain of any polynomial function is the set of **all** real numbers;

$D = \{x \mid x \in R\}$, and there is no restriction on the value of x.

The highest exponent (n) represents the degree of the function.

EXAMPLE

Find the degree of the factored function:

$$f(x) = x(2x + 3)(4x - 5)$$

Let's expand these brackets.

$$f(x) = x(2x + 3)(4x - 5) = (2x^2 + 3x)(4x + 5) = 8x^3 + 10x^2 + 12x^2 + 15x$$

$$f(x) = 8x^3 + 22x^2 + 15x$$

The degree of the function is 3

a. Power Functions

A power function is an expression of the form $f(x) = ax^n$ where a and n are real numbers.

The Cubic Function

$f(x) = x^3$

Domain $\{x | x \in R\}$

Range $\{y | y \in R\}$

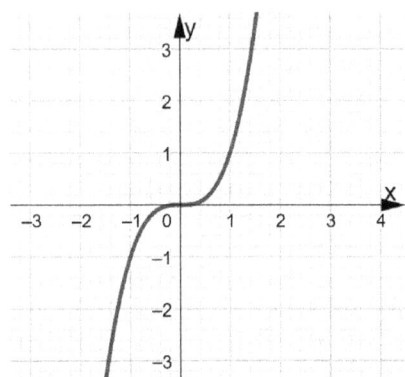

b. Concavity and Inflection Points

Concavity is used to describe the way a curve or in our case a graph of a function bends up or down.

Observation

1. When a line segment that is joining any two points on any curve is situated **completely below** that particular curve, - the curve is called **concave down** between the two points.

2. When a line segment that is joining any two points on any curve is situated **completely above** that particular curve, - the curve is called **concave up** between the two points.

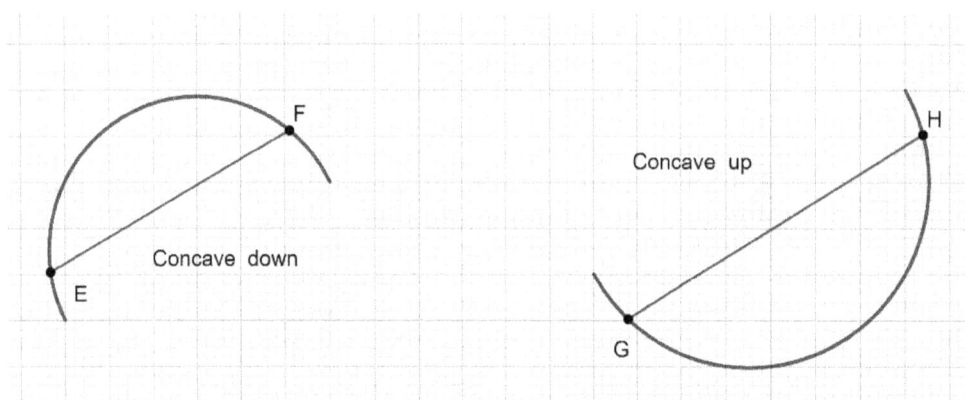

A **point of inflection** - a point where a graph of a function changes **concavity**.

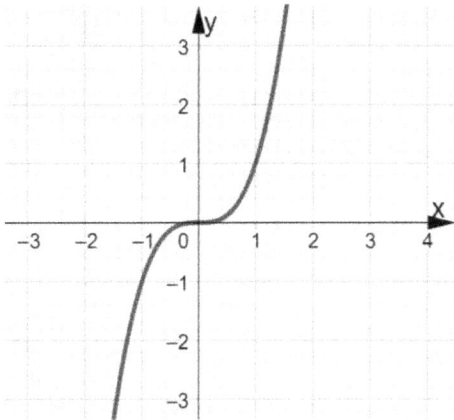

c. The Zeros (roots) of a Polynomial Function

We know that one way the zeros (roots) of a polynomial function can simply be identified when the function is in factored form.

For example, to determine the zeros of the given quadratic function

$y = 3x^2 - 14x + 8$

We simply factor this trinomial.

$y = 3x^2 - 14x - 8 = (3x - 2)(x - 4)$

We know as well that the product of any two factors equals zero when one of the factors is zero.

So, from $(3x - 2)(x - 4) = 0$ we have

$3x - 2 = 0 \Rightarrow x = \frac{2}{3} = 0.67$

Or,

$x - 4 = 0 \Rightarrow x = 4$

If we are graphing the function $f(x) = 3x^2 - 14x + 8$

Zeros are the x coordinates where the function becomes zero, or the x coordinates where the graph of the function intersects x axis.

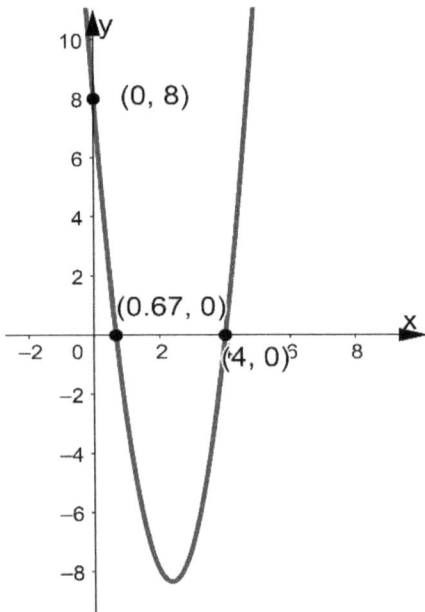

Y intercept is the constant of the general form of the function, or where x is zero.

PRACTICE

1) What is the degree of the function: $f(x) = x(x+3) - x^2 + 4$?

2) Analyze the concavity and the inflection point of $f(x) = x^3$

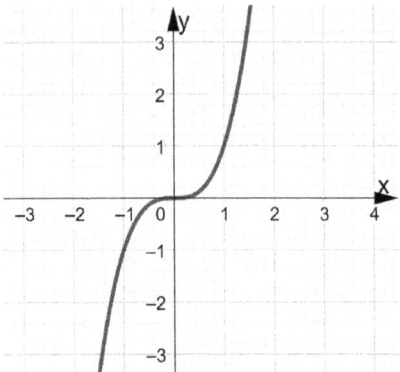

3) Find the zeros for the function $f(x) = 2x^2 + 3x - 4$

4) Find the zeros for the function $f(x) = x^2 + 2x - 3$

5) Find the zeros for the function $f(x) = x^2 + 2x + 7$

7.C. PIECEWISE FUNCTIONS

A *piecewise function* behaves differently on certain intervals of the Domain. It can be split in subsections.

EXAMPLE

$$f(x) = \begin{cases} x \text{ for } x \le -2 \\ 1 \text{ for } -2 < x < 3 \\ 3x - 2 \text{ for } x \ge 3 \end{cases}$$

This function has three subsections.

As we can see for all the values

$x \le -2 \ f(x) = x$.

For $-2 < x < 3$ the function is a horizontal line that crosses y axis at y=1.

For all the values of x greater or equal to 3, the function is a straight line with slope equal with 2 and

y-intercept equal with -2.

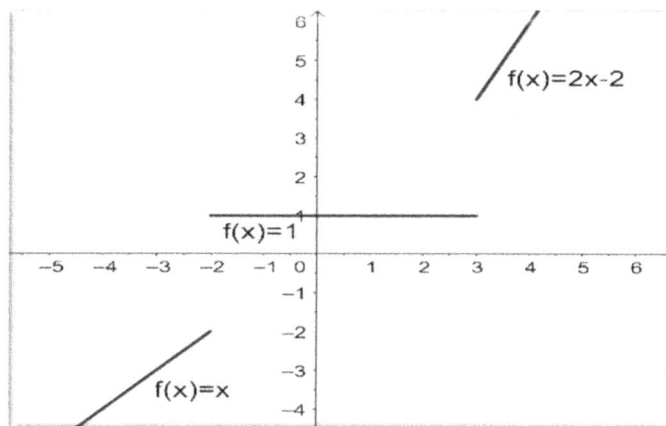

EXAMPLE

$$f(x) = \begin{cases} 2 - x \text{ for } x < 1 \\ 3 \text{ for } x \ge 1 \end{cases}$$

This function has two subsections.

As we can see for all the values

$x \le 1 \ f(x) = 2 - x$.

The straight line with slope of -1 crosses y axis at y=2.

For all the values of x greater or equal to 1, the function is a straight line with slope equal with 2.

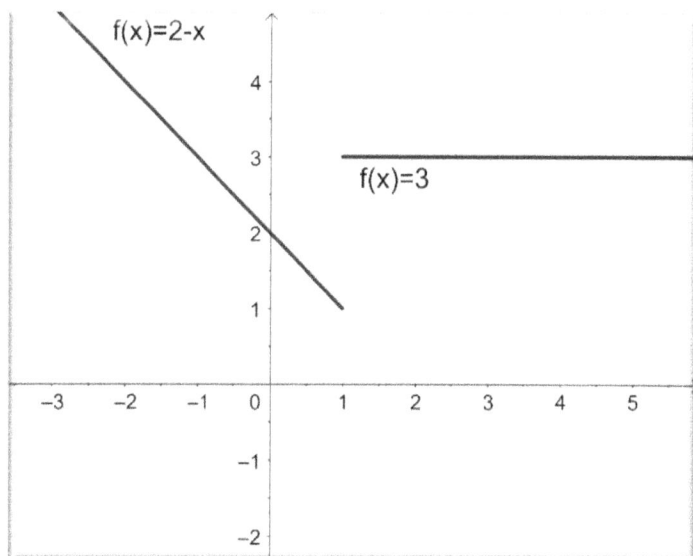

PRACTICE

Determine if the expressions below represent piecewise functions.

1) $f(x) = 2x^2 + 5x - 3, x \in R$

2) $f(x) = \begin{cases} 2x \ for \ x < 0 \\ 3 \ for \ x \geq 0 \end{cases}$

3) $f(x) = 4x + 3, x \in R$

4) $f(x) = 3\sin(x - 2) + 3, x \in R$

5) $f(x) = \sqrt{x - 1}$

7.D. TRIGONOMETRIC FUNCTIONS

The 6 trigonometric functions are:

$$\sin(x)\,;\cos(x)\,;\tan(x) = \frac{\sin(x)}{\cos(x)}\,;\cot(x) = \frac{\cos(x)}{\sin(x)} = \frac{1}{\tan(x)}\,;\sec(x) = \frac{1}{\cos(x)}\,;\csc(x) = \frac{1}{\sin(x)}$$

The trigonometric functions are _periodic_ functions. They repeat themselves periodically.

EXAMPLE

The graph of sin(x) is represented below.

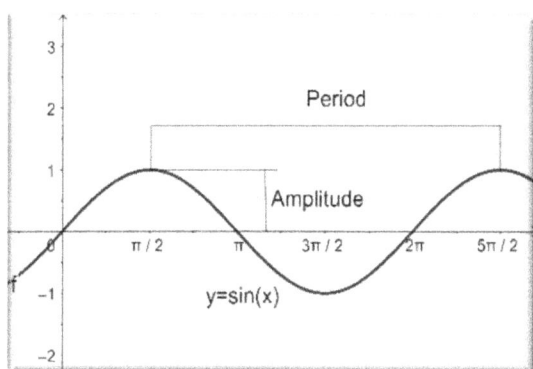

One <u>radian</u> is the measure of an angle subtended at the center of a circle by an arc which is equal in length to the radius of the circle as it can be seen in figure below.

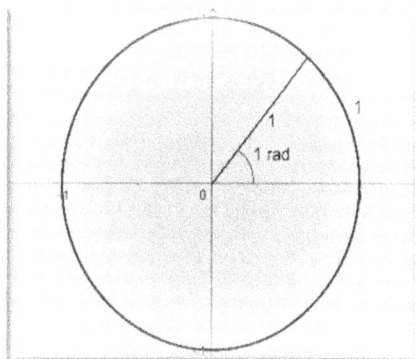

$$2\pi \; radians = 360^0 \, so, \pi \; radians = 180^0$$

As it can be seen below the period for

$$f(x) = \sin(x) \; and \; g(x) = \cos(x) \; is \; 2\pi, instead \; h(x) = \tan(x) \; has \; a \; period \; of \; \pi$$

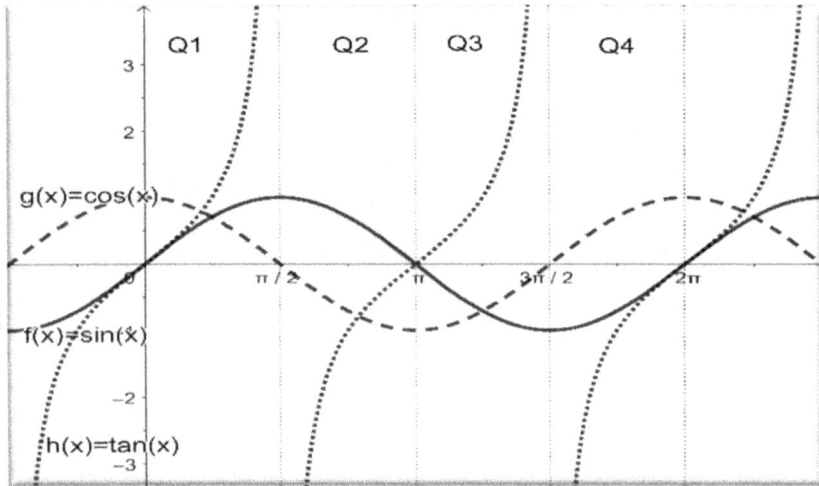

Q1 – first quadrant
Q2 – second quadrant
Q3 – third quadrant
Q4 – fourth quadrant

The sign of functions sin(x); cos(x) and tan(x) in each quadrant is shown below.

	Q1	Q2	Q3	Q4
Sin(x)	+	+	-	-
Cos(x)	+	-	-	+
Tan(x)	+	-	+	-

EXAMPLE

Tan(x) is negative in quadrants Q2 and Q4.

In a right-angle triangle, with $\sphericalangle\phi \neq 90^0$ we have the following _trigonometric ratios_:

$$\sin \sphericalangle\phi = \frac{opposite}{hypotenuse}$$

$$\cos \sphericalangle\phi = \frac{adjesant}{hypotenuse}$$

$$\tan \sphericalangle\phi = \frac{opposite}{adjesant}$$

EXAMPLE

In the right-angle triangle ΔABC $with$ $\sphericalangle B = 90^0$.

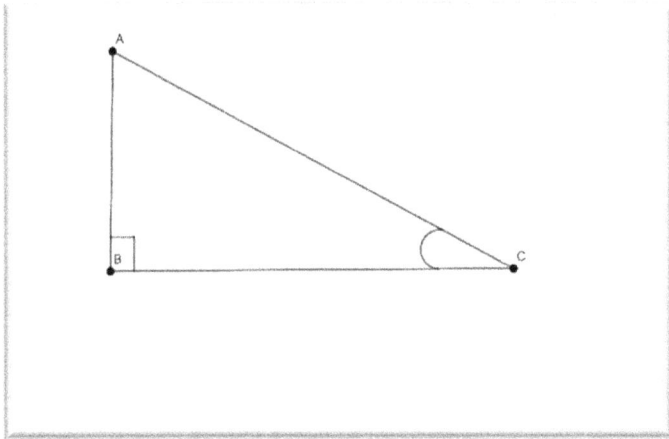

$$\sin \sphericalangle C = \frac{opposite}{hypotenuse} = \frac{AB}{AC}$$

$$\cos \sphericalangle C = \frac{adjesant}{hypotenuse} = \frac{BC}{AC}$$

$$\tan \sphericalangle C = \frac{opposite}{adjesant} = \frac{AB}{BC}$$

PRACTICE

Determine which statement is correct.

1) One radian is the measure of an angle subtended at the center of a circle by an ark which is equal in length to the radius of the circle.

2) $\pi \, radians = 360^0$

3) In Quadrant 1 $\sin (\alpha)$ is positive

4) If $\cos(\alpha_1) = k, k \geq 0$ the other value of α that is solution of the equation is: $\alpha_2 = 2\pi - \alpha_1$ (radians)

5) $\cos(\alpha) = \frac{opposite}{adjacent}$

7.E. GRAPHS OF TRIGONOMETRIC FUNCTIONS

a. Graphing sine and cosine functions

Sine and cosine trigonometric functions have the domain all the real values for x, and the range the interval [-1, 1].

EXAMPLE

The graph of $f(x) = \sin(x)$ is shown below.

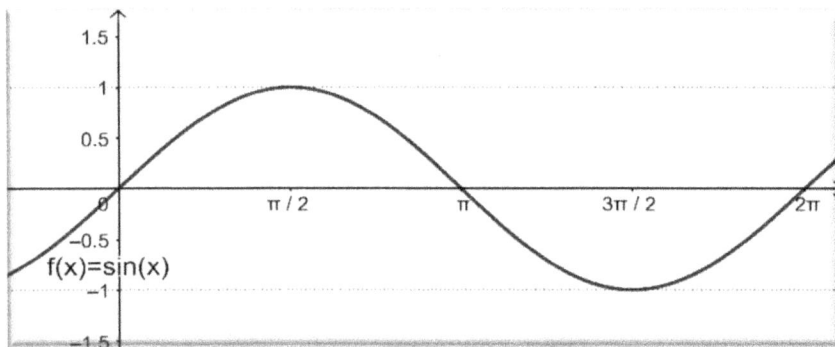

The value of $\sin\left(\frac{3\pi}{2}\right) = -1 \mp 2n\pi, n$ is integer

The value of $\sin(2\pi) = 0 \mp 2n\pi, n$ is integer

EXAMPLE

The graph of $f(x) = \cos(x)$ is shown below.

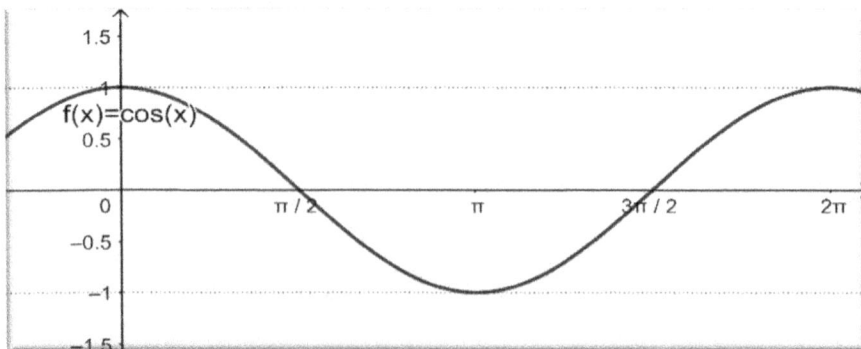

The value of $\cos\left(\frac{3\pi}{2}\right) = 0 \mp 2n\pi, n$ is integer

The value of $\sin(2\pi) = 1 \mp 2n\pi, n$ is integer

PRACTICE

Determine which statement is correct.

1) For each of values of α, the values of $\sin(\alpha)$ are:

α	0	$\dfrac{\pi}{6}$	$\dfrac{\pi}{2}$	$\dfrac{\pi}{3}$	π
$\sin(\alpha)$	0	$\dfrac{1}{2}$	1	$\dfrac{\sqrt{3}}{2}$	0

2) The graph of $\sin(\alpha)$ is:

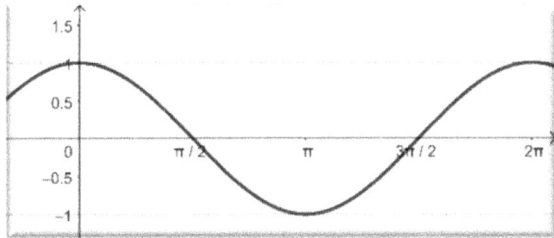

3) The minimum of $\cos(\alpha)$ is -1

4) The maximum of $\sin(\alpha)$ is -1

5) The graph of $\cos(\alpha)$ is:

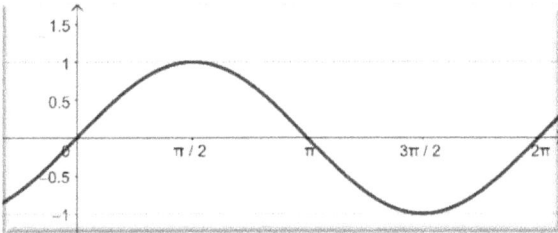

6) For each of values of α, the values of $\cos(\alpha)$ are:

α	0	$\dfrac{\pi}{6}$	$\dfrac{\pi}{2}$	$\dfrac{\pi}{3}$	π
$\cos(\alpha)$	1	$\dfrac{1}{2}$	2	$\dfrac{1}{2}$	0

b. Graphing tangent and cotangent functions

Tangent and cotangent trigonometric functions have the domain all the real values for x except the ones where either sine or cosine functions are zero.

It can be seen that for values of $x = \frac{\pi}{2} \mp k\pi, k - integer$ the tangent is undefined. For these values of x there are vertical asymptotes. The range includes all real numbers.

Asymptotes, are lines that graph of the functions go towards and the distance between the graph and asymptote approaches zero but never becomes zero.

The tangent graph is shown below.

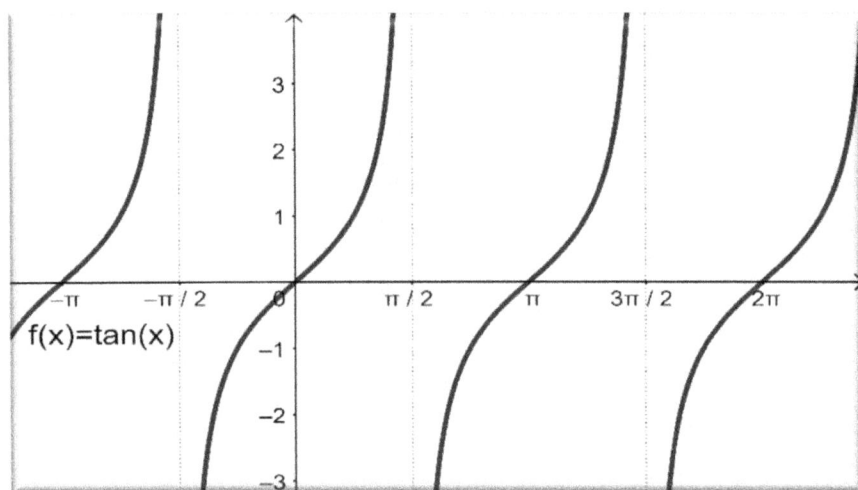

EXAMPLE

The tangent of zero is zero.

We know that $\tan(x) = \frac{\sin(x)}{\cos(x)}$ so, tangent function is zero when sine function is zero.

Tangent function is <u>undefined</u> when cosine function is zero.

EXAMPLE

Sine function is zero for $x = 0, and\ x = \mp k\pi\ where\ k - integer$.

We know that $\cot(x) = \frac{\cos(x)}{\sin(x)}$. The cotangent function is undefined for $x = \mp k\pi, k -$

integer where

The sign functions Sin(Φ), Cos(Φ), and Tan(Φ) in each quadrant is shown below.

	Q1	Q2	Q3	Q4
Sin(x)	+	+	-	-
Cos(x)	+	-	-	+
Tan(x)	+	-	+	-

The values for the special angles in a right-angle triangle are given below.

	30^0	60^0	45^0
Sin(Φ)	$\dfrac{1}{2}$	$\dfrac{\sqrt{3}}{2}$	$\dfrac{\sqrt{2}}{2}$
Cos(Φ)	$\dfrac{\sqrt{3}}{2}$	$\dfrac{1}{2}$	$\dfrac{\sqrt{2}}{2}$
Tan(Φ)	$\dfrac{\sqrt{3}}{3}$	$\sqrt{3}$	1

EXAMPLE

If $\sin(30^0) = \frac{1}{2}$ and, $\cos(30^0) = \frac{\sqrt{3}}{2}$ then:

$$\tan(30^0) = \frac{\sin(30^0)}{\cos(30^0)} = \frac{\frac{1}{2}}{\frac{\sqrt{3}}{2}} = \frac{1}{2} \div \frac{\sqrt{3}}{2} = \frac{1}{2} \times \frac{2}{\sqrt{3}} = \frac{1}{\sqrt{3}} = \frac{\sqrt{3}}{3}$$

The cotangent graph is shown below.

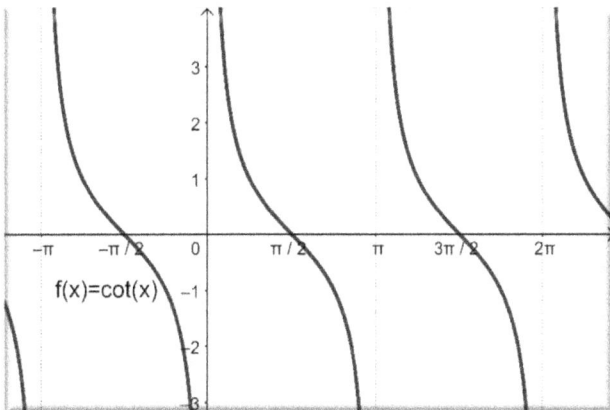

f(x)=cot(x)

It can be seen that for values of $x = \pi \mp k\pi, k - integer$ the cotangent is undefined. For these values of x there are vertical asymptotes.

EXAMPLE

If $\sin(30^0) = \frac{1}{2}$ and, $\cos(30^0) = \frac{\sqrt{3}}{2}$ then:

$$\cot(30^0) = \frac{\cos(30^0)}{\sin(30^0)} = \frac{\frac{\sqrt{3}}{2}}{\frac{1}{2}} = \frac{\sqrt{3}}{2} \div \frac{1}{2} = \frac{\sqrt{3}}{2} \times \frac{2}{1} = \sqrt{3}$$

If $\sin(60^0) = \frac{\sqrt{3}}{2}$ and, $\cos(60^0) = \frac{1}{2}$ then:

$$\cot(60^0) = \frac{\cos(60^0)}{\sin(60^0)} = \frac{\frac{1}{2}}{\frac{\sqrt{3}}{2}} = \frac{1}{2} \div \frac{\sqrt{3}}{2} = \frac{1}{2} \times \frac{2}{\sqrt{3}} = \frac{1}{\sqrt{3}} = \frac{\sqrt{3}}{3}$$

PRACTICE

Determine which statement is correct.

1) The tangent function is undefined at angles of 180^0 and 270^0 S

2) The tangent graph looks like the one below for $-\frac{\pi}{2} < \alpha < \frac{\pi}{2}$

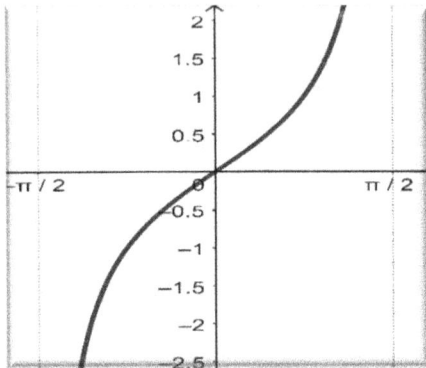

3) The cotangent function is zero for $\frac{\pi}{2} + k\pi, k$ $integer$

4) The tangent function is zero for $\frac{\pi}{2} + k\pi, k$ $integer$ A

5) For 60^0 the tangent is $\sqrt{3}$

6) Tangent of 45^0 is 1

7) Cotangent of 30^0 is 2

8) The tangent function has an amplitude of 100

9) Cotangent of 45^0 is 1

10) The formula of tangent in terms of sine and cosine is $\tan(\alpha) = \frac{\cos(\alpha)}{\sin(\alpha)}$

7.F. INVERSE TRIGONOMETRIC FUNCTIONS

Remember that an *inverse* function $f^{-1}(x)$ is the function that has the Domain equal with the Range of the original function $f(x)$. The Range of the inverse function equals the Domain of the original function.

EXAMPLE

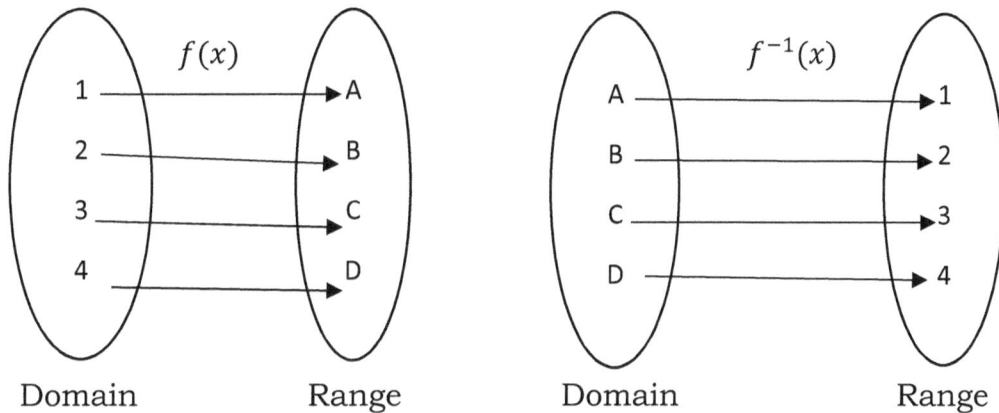

| Domain | Range | Domain | Range |

Because we are talking here about trigonometric functions like sine, cosine and tangent, in the notation for inverse trigonometric function,
we substitute $f^{-1}(x)$ with $\Phi = sin^{-1}(x)$, where:
x is value of the original trigonometric function.
Φ is the angle we are looking for.

EXAMPLE

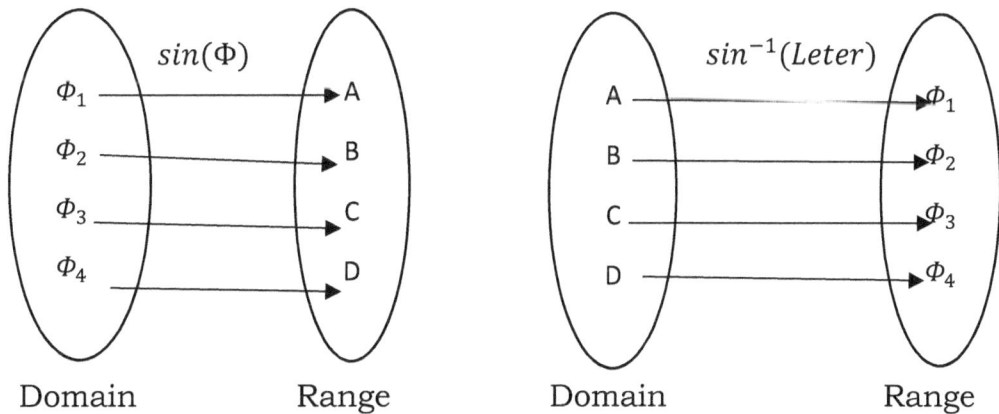

| Domain | Range | Domain | Range |

EXAMPLE

If $\sin(\Phi) = 0.345$, then $\Phi = sin^{-1}(0.345) = 20.18^0$

If $\tan(\Phi) = 1.89$, then $\Phi = tan^{-1}(1.89) = 62.11^0$

PRACTICE

Determine which statement is correct.

1) *Inverse trig functions* do the opposite of the "regular" trig functions.

2) The angle $\propto$ in right angle triangle PSQ is: 45^0

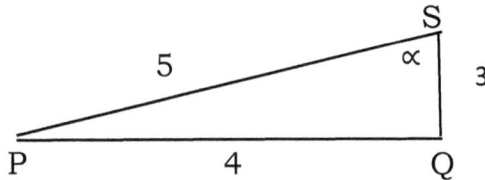

3) The angle $\propto$ in right angle triangle PSQ is: 35^0

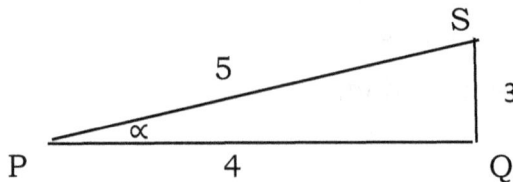

4) $Tan^{-1}(\sqrt{3})$ is 60^0

5) The domain of $f(x) = sin^{-1}(x)$ is $[0, 2\pi]$

7.G. GRAPHS OF INVERSE TRIGONOMETRIC FUNCTIONS

The inverse function of sine is $f^{-1}(x) = sin^{-1}(x) = \arcsin(x)$.
The graph of $f^{-1}(x) = sin^{-1}(x) = \arcsin(x)$ is shown below.

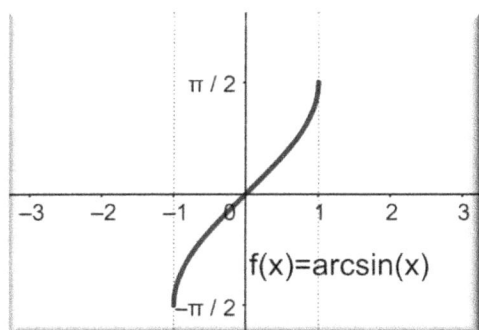

The domain is the interval [-1, 1].

The range is the interval $\left[-\frac{\pi}{2}, \frac{\pi}{2}\right]$.

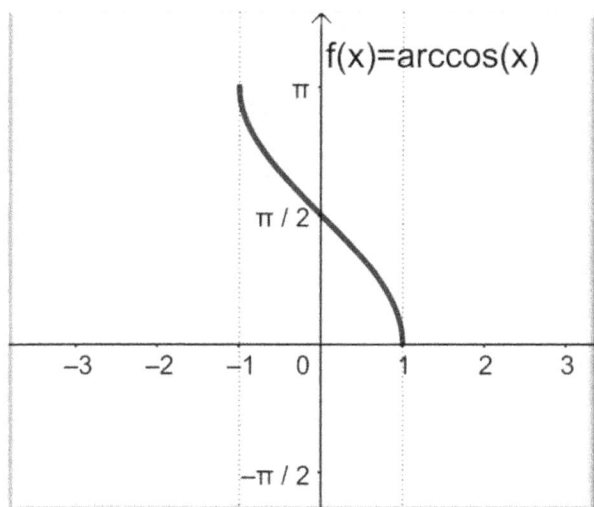

EXAMPLE

$f(0) = \arcsin(0) = 0$

$f(-1) = \arcsin(-1) = -\frac{\pi}{2}$

The graph of $f^{-1}(x) = cos^{-1}(x) = \arccos(x)$ is shown below.

The domain is the interval [-1, 1].
The range is the interval $[0, \pi]$.

EXAMPLE

$f(-1) = \arccos(-1) = \pi$

The graph of $f^{-1}(x) = tan^{-1}(x) = \arctan(x)$ is shown below.

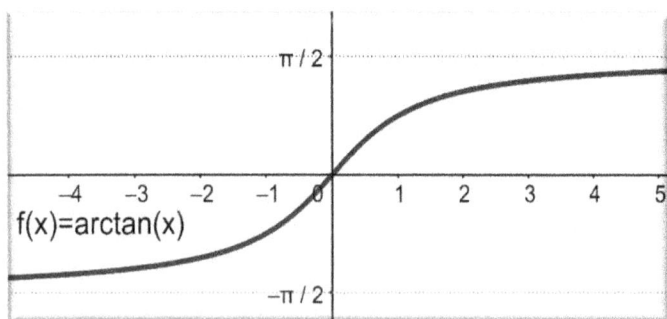

f(x)=arctan(x)

The domain is the real numbers.
The range is the interval $\left[-\frac{\pi}{2}, \frac{\pi}{2}\right]$.

EXAMPLE

$f(1) = \arctan(1) = \frac{\pi}{4}$

$f(-4) = \arctan(-4) = -0.47\,\pi$

The graph of $f^{-1}(x) = cot^{-1}(x) = \text{arccot}(x)$ is shown below.

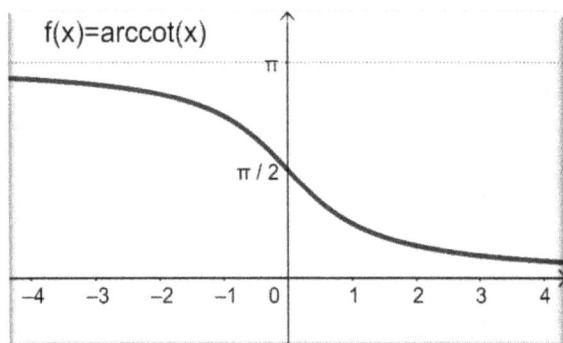

f(x)=arccot(x)

The domain is the real numbers.
The range is the interval $[0, \pi]$.

EXAMPLE

$f(-1) = \text{arccot}(-1) = 0.75\pi$

$f(-4) = \text{arccot}(-4) = 0.92\,\pi$

We could calculate the arccot(x) as $\frac{\pi}{2} - [\arctan(x)]$.

EXAMPLE

$f(-1) = \text{arccot}(-1) = \frac{\pi}{2} - [\arctan(x)] = \frac{\pi}{2} - [\arctan(-1)] = \frac{\pi}{2} - \left[-\frac{\pi}{4}\right] = \frac{\pi}{2} + \frac{\pi}{4} = \frac{3\pi}{4} = 0.75\pi$

$f(-4) = \text{arccot}(-4) = \frac{\pi}{2} - [\arctan(x)] = \frac{\pi}{2} - [\arctan(-4)] = \frac{\pi}{2} - [-0.42\pi] = 0.92\,\pi$

PRACTICE

Determine which statement is correct.

1) The domain of $f(x) = cos^{-1}(x)$ is [-1,1]

2) The value of $f^{-1}(1) = cos^{-1}(1)$ is 2

3) The value of $f(0) = cos^{-1}(0)$ is 1

7.H. INTRODUCTION TO LOGARITHMS

a. Definition, rules

We remember that $2^4 = 16$. It means that we multiply 2 by itself four times to get the result 16. Now, let's suppose we want to write the exponent 4 in terms of the result and the base.

This <u>exponent</u> will be written as the logarithm of base 2 from 16.

$4 = \log_2 16$

The logarithm is the exponent at which we have to rise the base to obtain the argument of the logarithm.

EXAMPLE

$\log_3 27 = 3$

Indeed rising 3 at exponent 3 the result is 27.

Rules of logarithms

1. Multiplication into addition

$\log_a(b \times c) = \log_a b + \log_a c$

EXAMPLE

$\log_4(5 \times 7) = \log_4 5 + \log_4 7$

2. Division into subtraction

$\log_a(b \div c) = \log_a b - \log_a c$

EXAMPLE

$\log_2(9 \div 3) = \log_2 9 - \log_2 3$

3. The argument at an exponent

$\log_a b^w = w \times \log_a b \quad b > 0$

EXAMPLE

$\log_5 7^9 = 9 \times \log_5 7$

4. Changing the base

$$\log_a b = \frac{\log_c b}{\log_c a} \quad a, b > 0$$

EXAMPLE

$$\log_5 7 = \frac{\log_2 7}{\log_2 5}$$

And now, let's put it all together.

Simplify:

$$\log_3 5 + \log_3 (x + 2) - \log_3 (x^2 + 4x + 4) = \frac{\log_3 5(x+2)}{\log_3 (x^2+4x+4)} = \frac{\log_3 5(x+2)}{\log_3 (x+2)^2}$$

Change the base to base 7.

$$\log_5 9 = \frac{\log_7 9}{\log_7 5}$$

PRACTICE

Determine which answer is correct.

1) The expression $\frac{1}{3}\log_6 a + 5\log_6 b - 7\log_6 c$ will become $\log_6 \left(\frac{a^{\frac{1}{3}} * b^5}{c^7}\right); a, b, c > 0$

2) The expression $\log_3 5^7$ will become $27 \times \log_3 5$

3) The expression $\log_3 63 + \log_3 5 - \log_3 35$ equals 3

4) The single logarithm of expression $3(\log_5 a + \log_5 b) - 2\log_5 b$ is $46 \log_5 a^3 b; a, b > 0$

5) The expression $\log_3 \left(\frac{a^2}{b^4}\right)$ in terms of $\log_3 a$ and $\log_3 b$ is $2\log_3 a - 42 \log_3 b; a, b > 0$

b. Exponential and logarithmic equations

<u>Exponential expression</u> is one where the variable x is at the exponent, like 3^x.
We will have to use the laws of logarithms and powers to be able to solve exponential and logarithmic equations.

EXAMPLE

Solve for x:

a. $5^x = 125$

$5^x = 5^3$

$x = 3$

b. $21^{3x-4} = 441^{5x+3}$

$21^{3x-4} = (21^2)^{5x+3}$

$21^{3x-4} = 21^{2(5x+3)}$

$3x + 4 = 2(5x + 3)$

$3x + 4 = 10x + 6$

$-7x = 2$

$x = -\frac{2}{7}$

c. $\log_3 x = \log_3 27 - \log_3 6 , x > 0$

$\log_3 x = \log_3 \frac{27}{6} = \log_3 \frac{9 \times 3}{2 \times 3} = \log_3 \frac{9}{2}$

$x = \frac{9}{2} = 4.5$

d. $\log_3(2x - 1) + \log_3 2 = \log_3 x , \ 2x - 1 > 0, \ x > 0$

$\log_3 2(2x - 1) = \log_3 x$

$2(2x - 1) = x$

$4x - 2 = x$

$3x = 2$

$x = \frac{2}{3}$

Remember the natural logarithm written $\ln x$ where the base is e or Euler's number e = **2.71828.....**

PRACTICE

Determine which answer is correct.

1) The solution of the equation $620 = 2 * 3^{x+1}$ is $x = \log_3 310 - 1$

2) The solution of the equation $5 = \log_5 x + \log_5(x - 3)$ is $x = 36.54; x > 3$

3) The solution of the equation $\log_5(3x + 9) - \log_5(x + 3) = \log_5(x - 4)$ is $x = 37; x > 4$

4) The solution of the equation $3^{x+3} = 7^{2x+5}$ is $x = -0.953$

5) The solution of the equation $\log(x^2 + x - 6) - \log(x + 3) = 1$ is $x = 192$

7.I. EXPONENTIAL AND LOGARITHMIC FUNCTIONS

The general expression for an exponential function is $f(x) = a^x$, where a is a positive real number not equal with 1.

Let us graph the exponential function

$$f(x) = y = 3^x$$

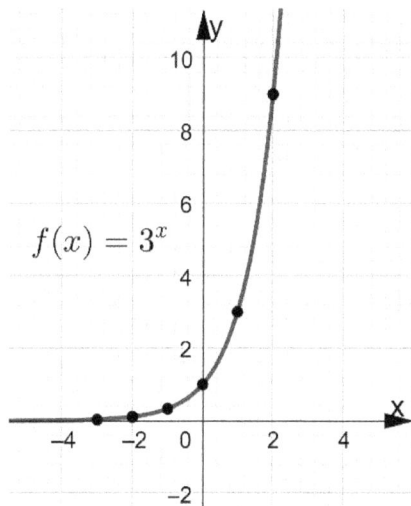

$f(x) = 3^x$

The function $f(x) = 3^x$ is represented here

This shape of the graph of the exponential function from down left to right up happens for the base of the power being bigger than 1.

For the constant a between 0 and 1, and for negative exponent the shape is up left to down right.

Let us represent the graph of $f(x) = 0.5^x$

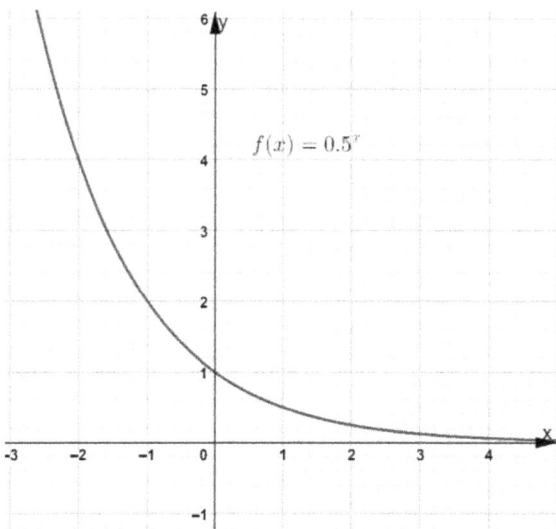

$f(x) = 0.5^x$

The domain of the exponential function

$$Domain = \{x|\, x \in R\}$$

The range of the exponential function is bigger than 0

$$Range = \{y|y > 0, y \in R\}$$

Remember, the logarithm function is the inverse of exponential function.

Remember what happens with the coordinates of an inverse function.

An <u>inverse</u> function represented by $f^{-1}(x)$ is the function that has the Domain equal with the Range of the original function $f(x)$. The Range of the inverse function equals the Domain of the original function.

$y = 3^x$ (x,y)	$\log_3 y = x$ (x,y)
$(-3, \frac{1}{27})$	$(\frac{1}{27}, -3)$
$(-2, \frac{1}{9})$	$(\frac{1}{9}, -2)$
$(-1, \frac{1}{3})$	$(\frac{1}{3}, -1)$
$(0,1)$	$(1,0)$
$(1,3)$	$(3,1)$
$(2,9)$	$(9,2)$

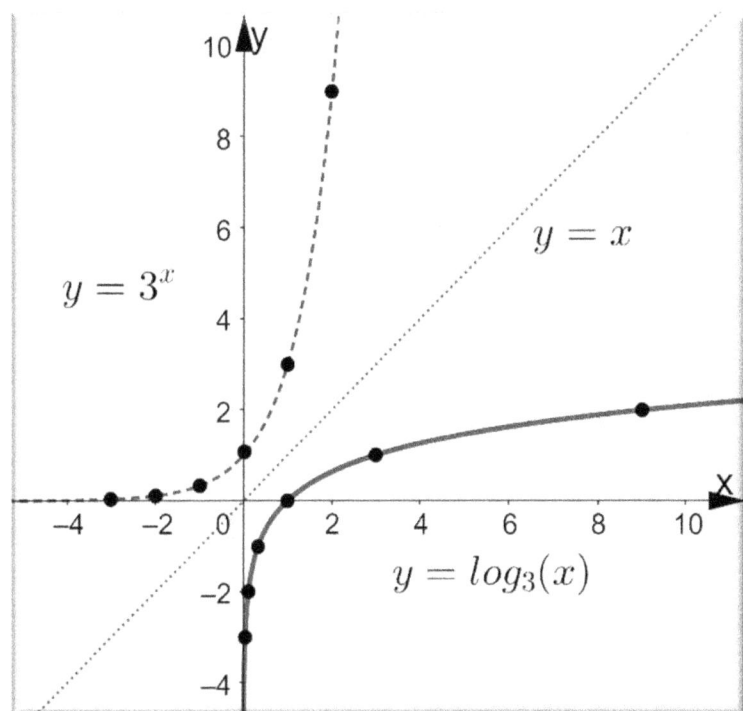

The domain of the exponential function is the range of the logarithmic function, the set of all real numbers.

The range of the exponential function is the domain of the logarithmic function, the set of all real values greater than 0.

PRACTICE

1) Represent the exponential function $f(x) = 4^x$

2) Represent the logarithmic function $f(x) = \log_2 x$.

3) Represent the exponential functions $f(x) = 2^x$ and $f(x) = (\frac{1}{2})^x$ on the same graph. What do you notice?

4) Represent the logarithmic functions $f(x) = \log_2 x$ and $\log_{\frac{1}{2}} x$. What do you notice?

5) Represent the exponential function $f(x) = 5^x$ and the logarithmic function $\log_5 x$ on the same graph. What do you notice?

CHAPTER 8

Transformations of Functions

8.A. INTRODUCTION TO TRANSFORMATIONS

In general, a **transformation** changes one or more of the location, shape, size, or orientation of an object in the coordinate plane.

The object could be, for example, a function like $y = x^2$, or a geometric shape, like $\triangle ABD$.

The initial object is called the **pre-image**.

The object that results after transforming the pre-image is called the **image**.

When moving from A to B to C in $\triangle ABC$, we would say that there is a clockwise orientation.

Let's create an image by sliding $\triangle ABC$ 5 units to the right and 2 units down.

What is $\triangle DEH$ in this case?

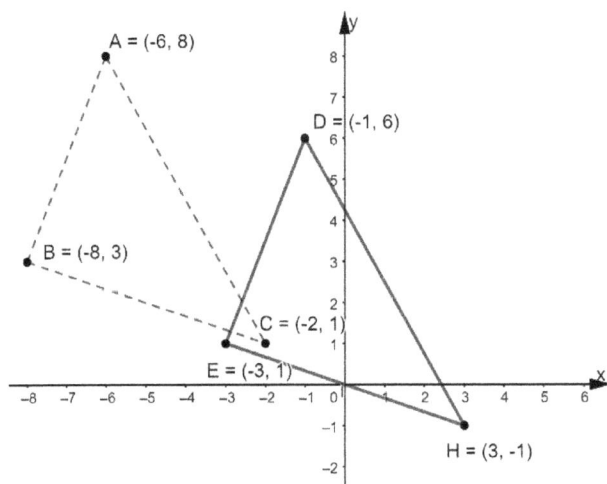

Pre-image		Image
(x,y)	$\rightarrow$	(x′,y′)
A (-6,8)	$\rightarrow$	D (−1,6)
B (-8,3)	$\rightarrow$	E (-3,1)
C (−2,1)	$\rightarrow$	H (3,−1)

We can describe the translation using mapping notation as

$(x,y) \rightarrow (x-5, y-2)$.

8.B. VERTICAL AND HORIZONTAL TRANSLATIONS

a. Vertical Translations

When the function is $y = f(x) + d$

1) When d>0, the graph of $y = f(x)$ is translated up d units
2) When d<0, the graph of $y = f(x)$ is translated down |d| units

EXAMPLE

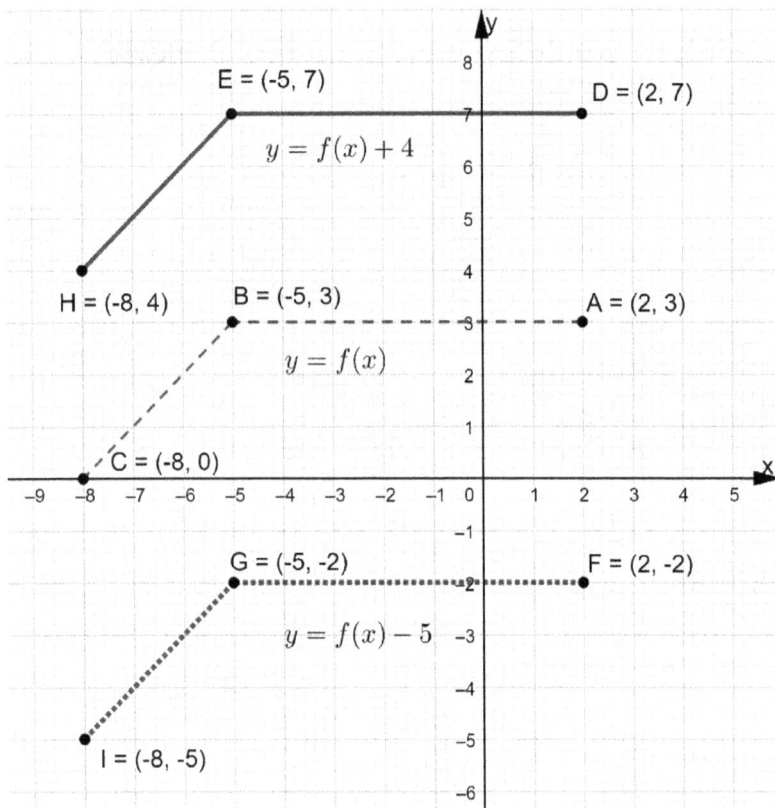

b. Horizontal Translations

When the function is $y = f(x - d)$

    1) When d>0, the graph of $y = f(x)$ is translated right d units
    2) When d<0, the graph of $y = f(x)$ is translated left $|d|$ units

EXAMPLE

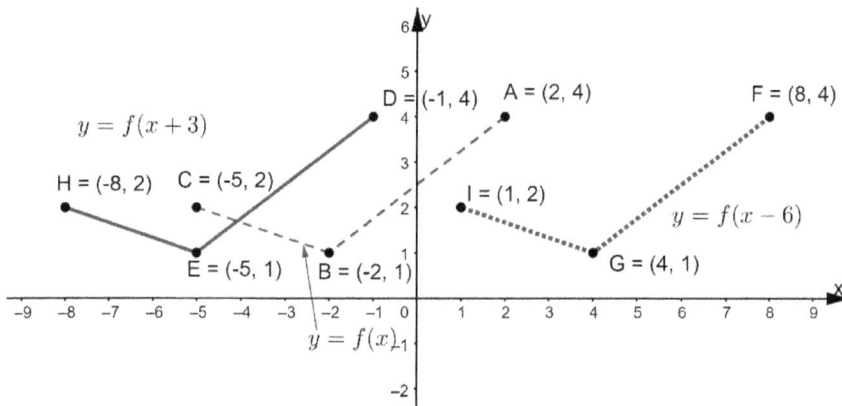

The **combined** formula will be:

$G(x) = f(x - d) + e$

EXAMPLE

Given the function f(x)shown below, sketch y=f(x+2)−3.

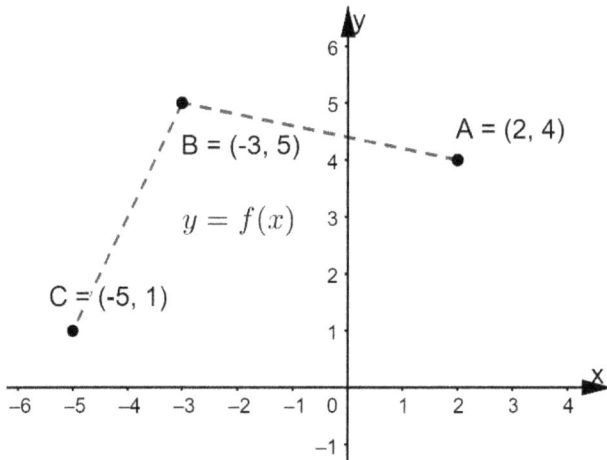

Comparing y=f(x+2)−3 to y=f(x−d)+e, we see that d=-2 and e=-3.
Therefore, the translation maps each point on y=f(x) 2 units to the left and 3 units down.

In mapping notation, the translation is written

$$(x,y) \rightarrow (x+2, y-3)$$

Here we have both pre image and the image.

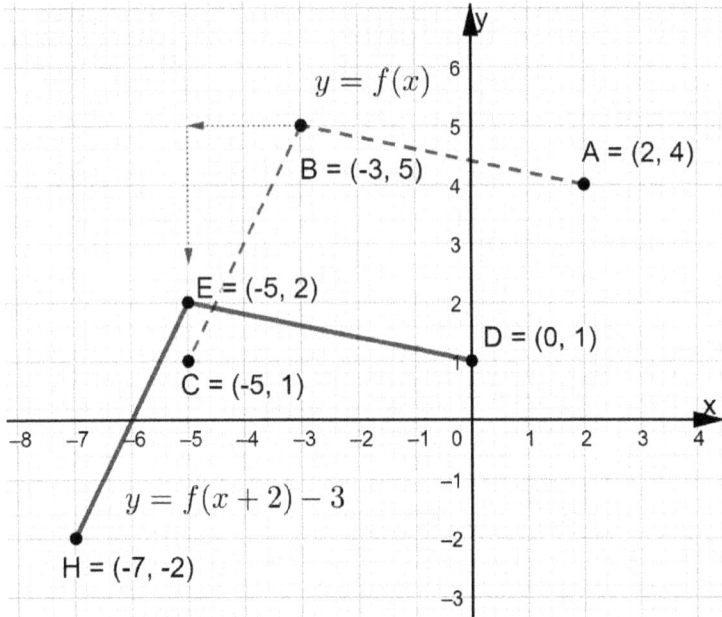

PRACTICE

Translate the graph below.

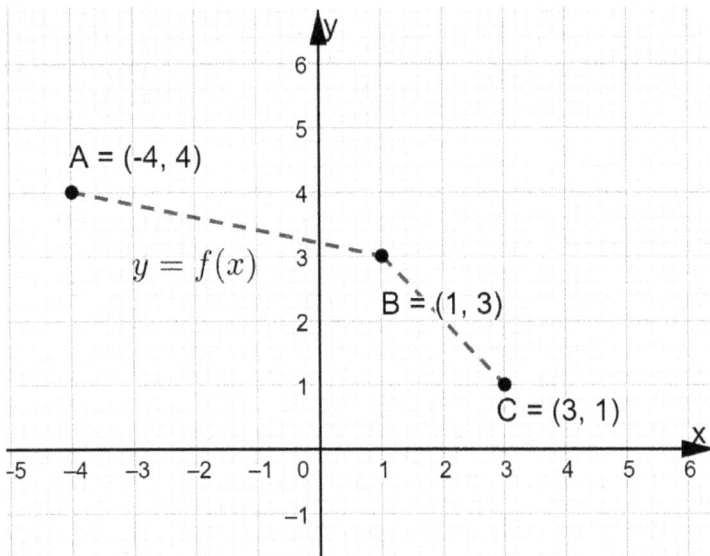

1) 3 units up

2) 2 units to the left

3) 1 unit down and 3 units to the right

4) 3 units up 2 units to the left

5) 2 units down and 4 units to the right

8.C. REFLECTIONS

On the graph below, $\triangle ABC$ is reflected in the y-axis. The image is the triangle DEH

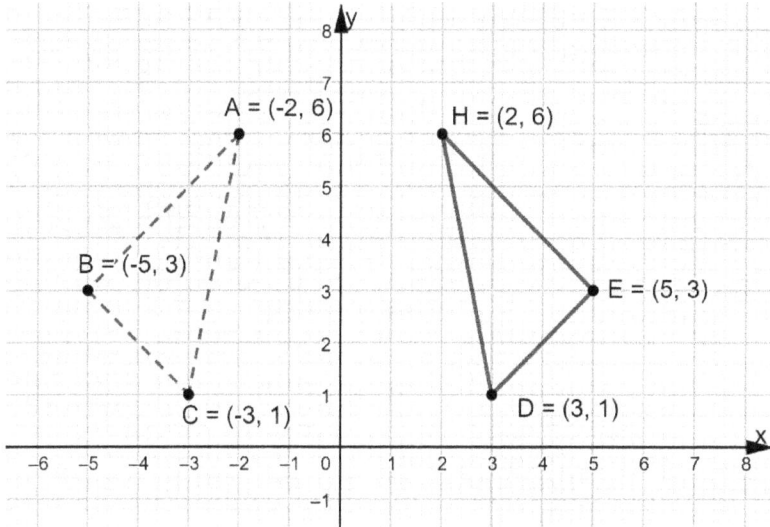

The **x coordinates** of all the points of the original figure **changes sign** into the opposite one.

A(-2,6) becomes H(2,6)

The **y coordinates** of all the points of the original figure will **remain the same** in the case of a reflection by y axis.

We can see that the size and shape of the triangle or a function doesn't change when reflecting it.

The line of the reflection is the right bisector of the line segment connecting any pre-image point (say A) with a point on the image (in this case H)

In the graph above the line of reflection is the y axis.

Vertical and horizontal reflections

Here we have a reflection in the x axis.

G(x) is the reflected function in x axis.

What do we notice?

In this case, the x coordinates remain the same.

The y coordinates change sign.

B(-4,4) becomes H(-4,-4)

A(4,4) becomes E(4,-4)

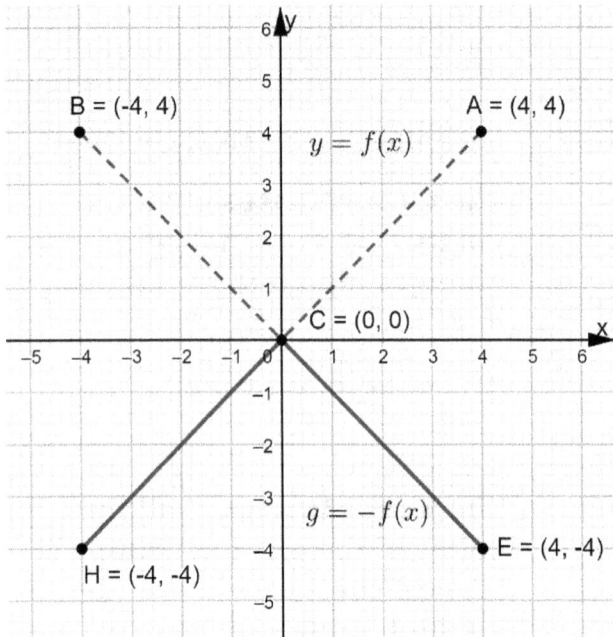

In general, if $g(x) = -f(x)$, then g(x) represents a reflection of f(x) in the x-axis.

Here we have a reflection in the y axis. The function g(x) is the reflected function.

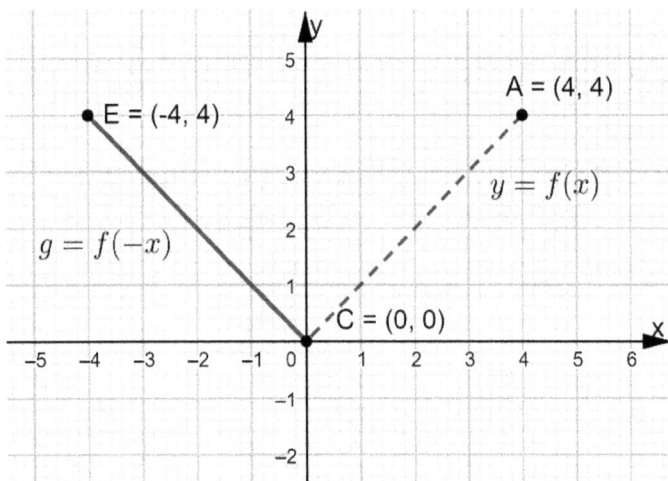

What do we notice?

The y coordinates remain the same.

The x coordinates change sign.

A(4,4) becomes E(-4,4)

In general, if g(x)=f(-x), it results that g(x) is a reflection of f(x) in the y axis.

PRACTICE

Reflect the graph below.

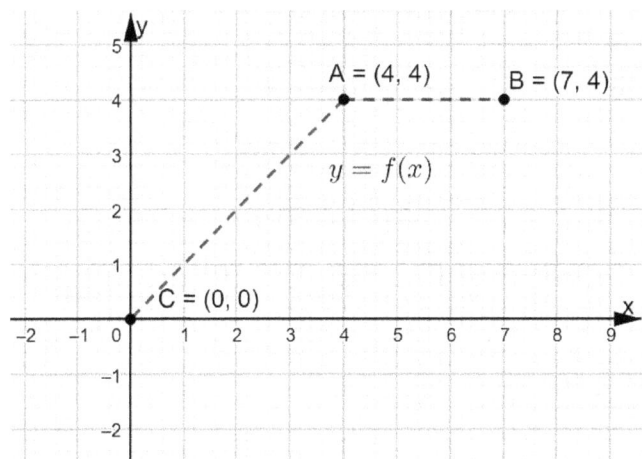

1) On x axis

2) On y axis

3) On x and y axis

NOTE

Even Functions

When a function is reflected in the y-axis and the y-axis is also its axis of symmetry, the reflection will leave the original function unchanged.

If $f(x) = x^2$ is reflected in the y axis, (its axis of symmetry) then the resulting image is:

$g(x) = f(-x) = (-x)^2 = x^2 = f(x)$

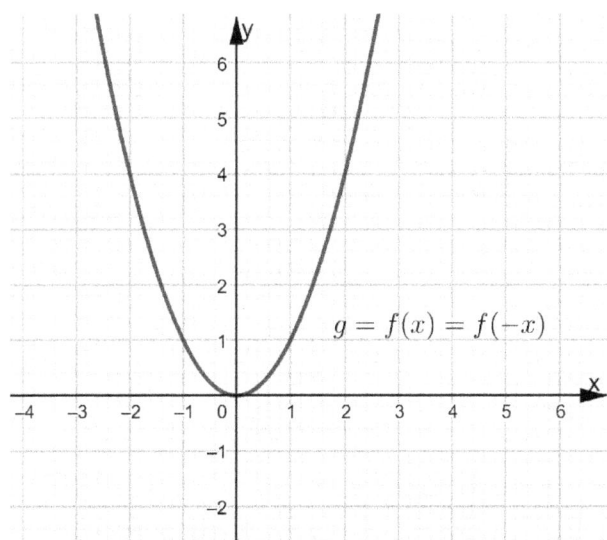

So, If $f(x) = f(-x)$ for all the values in the domain the function is aid to be **even**.

More formally, an even function is symmetrical about the y -axis and will map onto itself when reflected in the y -axis.

Odd Functions

Next, we will use the reciprocal function $f(x) = \frac{1}{x}$ to introduce the concept of odd functions.

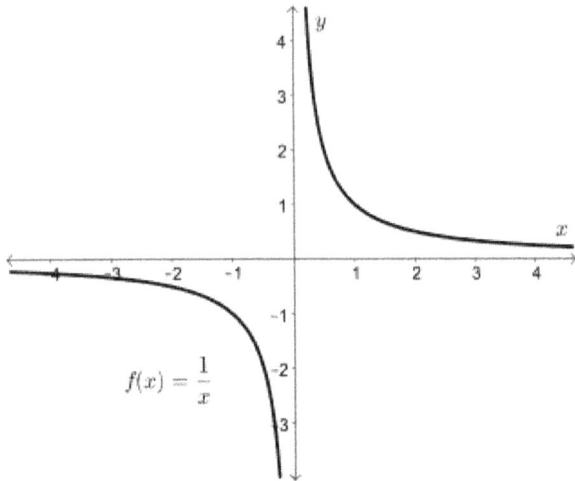

Now, let's reflect g(x) in y axis.

The reflection of g(x) in y axis will be

$$H(x) = g(-x) = \frac{1}{-x} = -\frac{1}{x} = f(x)$$

In general, when $f(-x) = -f(x)$ for all values of x in the domain, the function is called an **odd function.**

8.D. VERTICAL EXPANSION OR COMPRESSION

a. Transformation of a parabola

The general formula for the transformation of a parabola is:

$y = a(x + h)^2 + k$

Where:

a - A vertical stretch (or compression) by a factor determined from the size of a

When -1>a>1 there will be a vertical stretch or expansion

When -1<a<1 there will be a vertical stretch or compression

When a<0 there will be a reflection in the x axis.

k - A vertical translation (up or down) given by the size of k.

h - A horizontal translation (right or left) given by the size of h

In transformations from $y = x^2$ to obtain $y = a(x - h)^2 + k$ vertical stretches or compressions, and reflections in the x-axis, must take place **before vertical translations**.

It does not matter when the horizontal translation is applied.

EXAMPLE

Sketch $f(x) = -2(x - 3)^2 + 4$

We start with the reflection on x axis

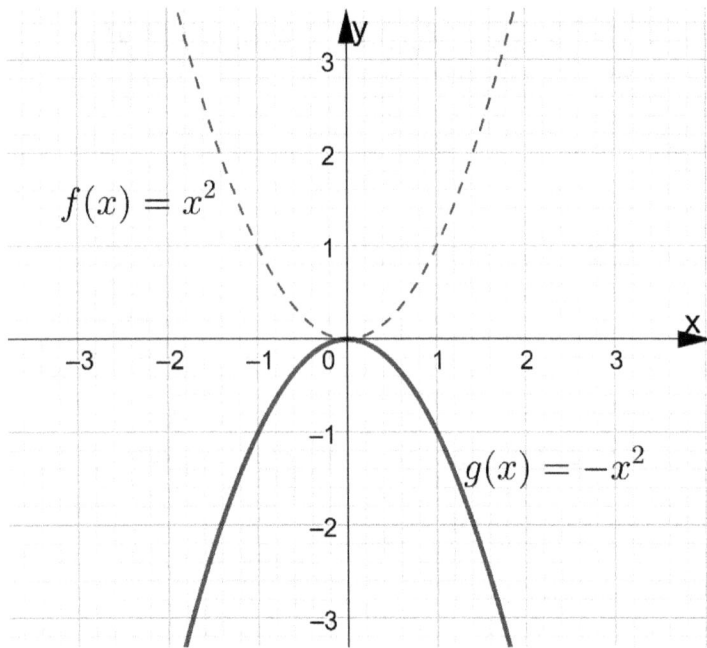

$f(x) = x^2$

$g(x) = -x^2$

Then we have the vertical expansion in y by a factor of 2

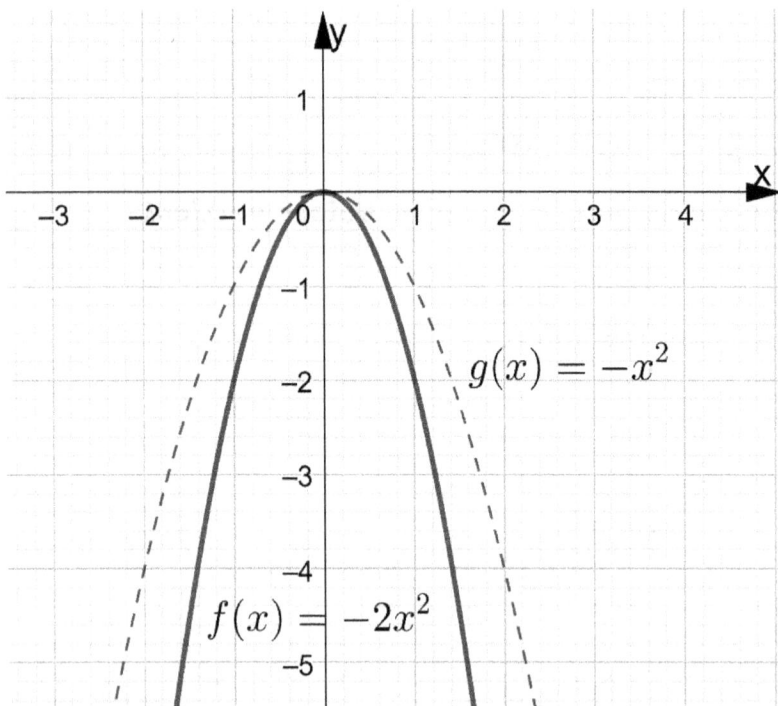

$g(x) = -x^2$

$f(x) = -2x^2$

Next is the translation 3 units to the right

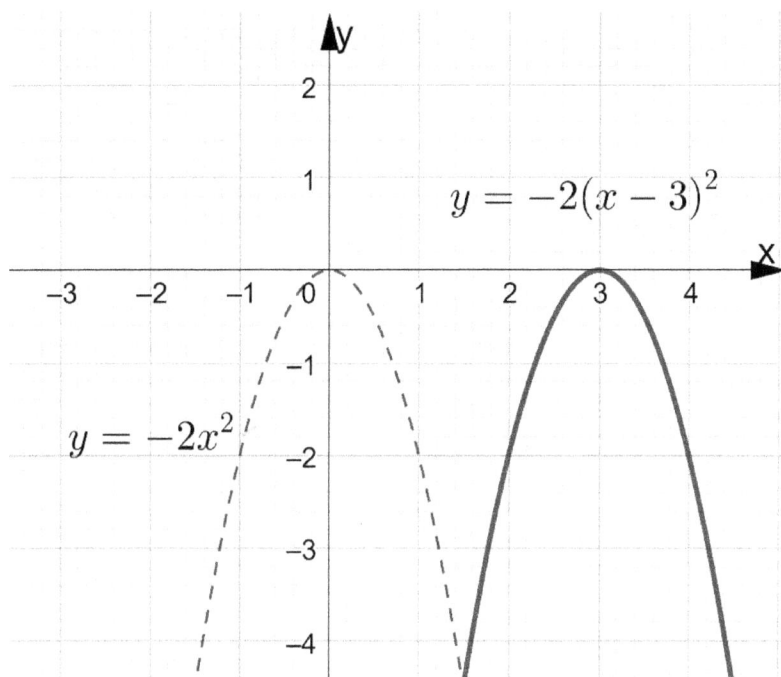

$$y = -2(x - 3)^2$$

$$y = -2x^2$$

The last step is the translation 4 units up

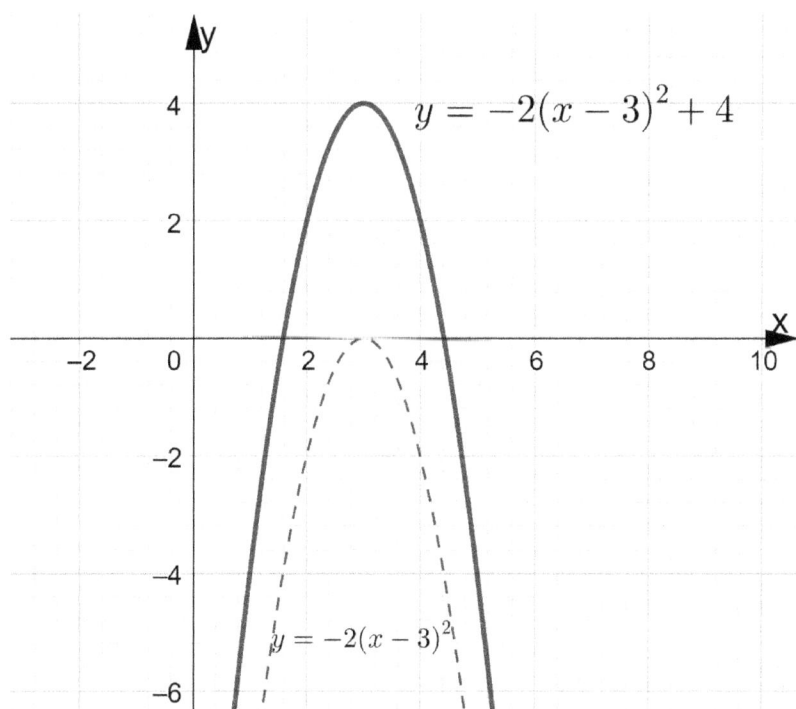

$$y = -2(x - 3)^2 + 4$$

$$y = -2(x - 3)^2$$

PRACTICE

1) Sketch $h(x) = -\frac{1}{2}(x + 2)^2 + 3$

2) Sketch $y = h(x) = -2(x - 4)^2 + 1$

3) Sketch $y = h(x) = 3(x + 2)^2 - 4$

4) Sketch $y = h(x) = 3(x + 2)^2 - 4$

5) Sketch $y = h(x) = \frac{1}{3}(x + 3)^2 + 4$

CHAPTER 9

Systems of Equations

9.A. SYSTEM OF LINEAR EQUATIONS USING TWO EQUATIONS AND TWO UNKNOWNS.

Let us have the equation

$3x - y = 2$

If we graph it, it looks like this:

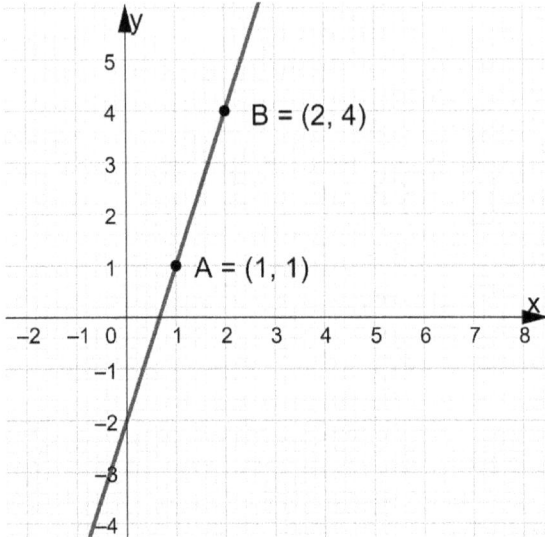

It has millions of points that will satisfy the equation $3x - y = 2$

If we take point $(1,1)$ we will have:

$3(1) - 1 = 2$

3-1=2 so, 2=2

If we take point $(2,4)$ we will have:

$3(2) - 4 = 2$

6-4=2

2=2

If we consider the line $y = -x + 2$ we will have the graph:

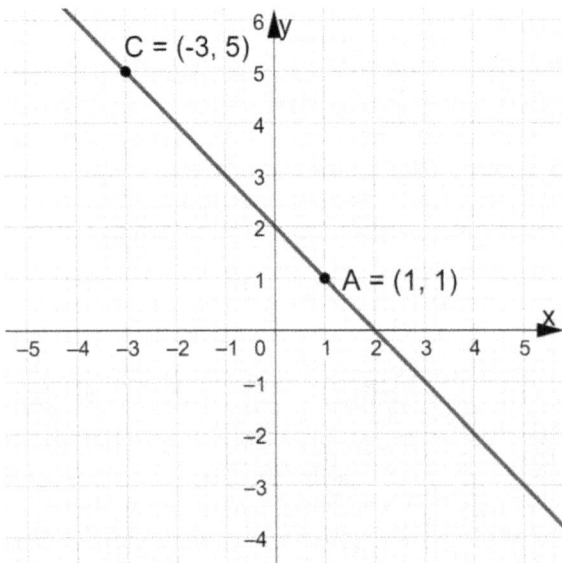

Again, it has an infinite number of points that will satisfy the equation $y = -x + 2$

If we take point $(-3,5)$ we will have

$5 = -(-3) + 2$

5=3+2

5=5

If we take point $(1,1)$ we will have:

$1 = -(1) + 2$

1=1

If we graph these two lines on the same graph there will be a point that will satisfy both equations

Here is the graph

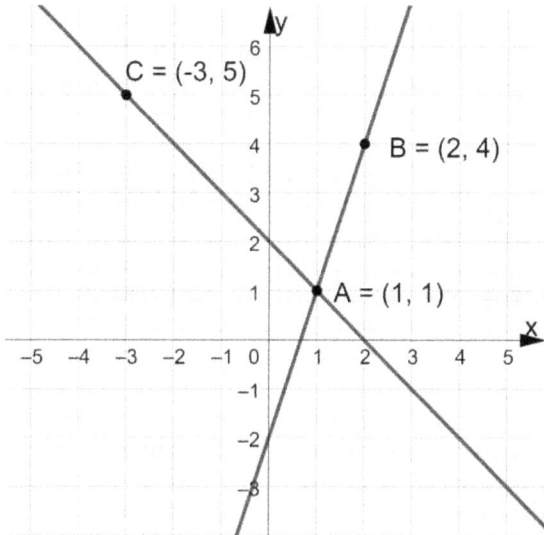

The solution of the system of two linear equation is the point, or the pair of coordinates x and y that satisfy both equations.

Point (1,1) is already checked before in both equations.

a. Solving the system of linear equations algebraically

Remember that when we have

3=3

6=6

If we add vertically, we would have

3+6=3+6

Or,

9=9

Substitution in an expression

If we consider the expression a+b+c, and we are given a=−3, b=2, and c=8, we can **substitute** these numerical values in for a, b, and c, and evaluate as follows:

$$-3+2+8=7$$

If $a = x - 5$, $b = x + 3$, and $c = 4 - x$, we can substitute these expressions in for a, b, and c, and simplify to:

$$a + b + c = x - 5 + x + 3 + 4 - x = x + 2$$

When we are asked to solve a system of equations, we are being asked to find values of the variables that satisfy all of the equations simultaneously.

To solve a system of equations having n variables, we typically need n equations written in those variables.

We will analyze two methods that help us to solve systems of linear equations. These methods are "by substitution" and "by elimination"

b. Solving the system of equations by substitution

First step: We start by isolating one variable in terms of the second variable. If it is possible, we choose a variable that has a coefficient of 1 to make your algebra more friendly.

Second step: Then we substitute the expression into the equation not used in the previous step. This gives us one equation with one unknown variable.

Third step: We then solve for the variable.

Fourth step: We substitute the value to solve for the second variable. In this step it does not matter which equation we choose to work with.

Fifth step: Check the solution in the equation not used in Step 4.

EXAMPLE

Solve the system of equations by substitution:

$$\begin{cases} 2x + 4y = -7 \\ x + 3y = -2 \end{cases}$$

First step: We start by isolating one variable in terms of the second variable. If it is possible, we choose a variable that has a coefficient of 1 to make your algebra more friendly.

In the second equation, we are writing x in terms of y
$x = -3y - 2$

Second step: Then we substitute the expression into the equation not used in the previous step. This gives us one equation with one unknown variable.
We substitute the expression of x into the first equation.

$2(-3y - 2) + 4y = -7$

Third step: We then solve for the variable.

$-6y - 4 + 4y = -7$
$-2y - 4 = -7$
$-2y = -3$
$y = \dfrac{-3}{-2} = \dfrac{3}{2} = 1.5$

Fourth step: We substitute the value to solve for the second variable. In this step it does not matter which equation we choose to work with.
We will substitute the value for y in the second equation.

$x + 3y = -2$

So,
$x + 3(1.5) = -2$

$x = -2 - 4.5 = -6.5$

Fifth step: Check the solution in the equation not used in Step 4.

$2x + 4y = -7$

$2(-6.5) + 4(1.5) = -7$

$-13 + 6 = -7$

$-7 = -7$

The solution is x=-6.5 and y=1.5.

These are the coordinates of the point of intersection between the two lines represented by the two linear equations.

$$\begin{cases} 2x + 4y = -7 \\ x + 3y = -2 \end{cases}$$

PRACTICE

1) Find the solution of the system of equations and check

$$\begin{cases} y = -2x + 1 \\ x - y = 2 \end{cases}$$

2) Find the solution of the system of equations

$$\begin{cases} 3x + 4y = -5 \\ 2x - 2y = -4 \end{cases}$$

3) Find the solution of the system of equations

$$\begin{cases} 3x + 3y = 4 \\ x - 2y = 3 \end{cases}$$

4) Find the solution of the system of equations

$$\begin{cases} 3x + 2y = 1 \\ 3x - y = 2 \end{cases}$$

5) Find the solution of the system of equations

$$\begin{cases} \frac{y+3}{1-2x} = \frac{1}{4} \\ x + 3y = 2 \end{cases}$$

c. Solving a system of linear equations by Method of Elimination

First step. If required, we multiply one or both equations by a factor to ensure that both equations have either an x or y coefficient that is either equal or opposite to each other.

Second step. We eliminate one variable by adding or subtracting the two equations. If the coefficients from Step 1 have the same sign we will subtract the equations, and if they have opposite signs, we will add the equations.

Third step. We then solve for the variable (say x)

Fourth step. We then substitute the value to solve for the second variable (say y).In this step it does not matter which equation we choose to work with.

Fifth step. We finally check the solution in the equation not used in Step 4

EXAMPLE

Solve the system of equations using elimination:

$$\begin{cases} x + 2y = 3 \\ 2x - 3y = 5 \end{cases}$$

First step. We multiply the first equation by a factor of -2 to ensure that both equations have an x coefficient that is either equal or opposite to each other.

$$\begin{cases} -2x - 4y = -6 \\ 2x - 3y = 5 \end{cases}$$

Second step. We eliminate one variable, here x, by adding vertically.

$$\begin{cases} -2x - 4y = -6 \\ 2x - 3y = 5 \end{cases}$$
$$0x - 7y = -1$$

Third step. We then solve for the variable y in this case.

$-7y = -1$

We divide both sides by -7

$y = \dfrac{1}{7} = 0.14$

PRACTICE

1) Solve the system of equations using elimination:
$$\begin{cases} y = 5x - 6 \\ y = 3x + 4 \end{cases}$$

2) Solve the system of equations using elimination:
$$\begin{cases} 2x + 3y = 4 \\ x - 2y = 3 \end{cases}$$

3) Solve the system of equations using elimination:
$$\begin{cases} \dfrac{y+2}{1-2x} = \dfrac{1}{3} \\ x + 3y = 2 \end{cases}$$

4) Solve the system of equations using elimination:
$$\begin{cases} \dfrac{5}{2}x - \dfrac{2}{3}y = 11 \\ 5x - 3y = 2 \end{cases}$$

5) Solve the system of equations using elimination:

$$\begin{cases} \dfrac{1}{2}x + \dfrac{2}{5}y = 1 \\ \dfrac{5}{6}x - \dfrac{1}{2}y = 2 \end{cases}$$

STEP BY STEP SOLUTIONS

1.B. introduction to powers

a. Definition, rules

Determine which answer is correct

1) The simplified expression of $\frac{5x^6y^9}{15x^4y^3}$ is $\frac{1}{3}x^2y^6; x, y \neq 0$

2) The simplified expression of $(-3a^2b^3)^3$ is $-27a^6; a \neq 0$

3) The simplified expression of $(\frac{-7x^4y^3}{xy})^2(\frac{x^2yz}{7x^2z})^3 = (-7x^3y^2)^2(\frac{y}{7})^3 = 49x^6y^4 * \frac{y^3}{49*7} = \frac{1}{7}x^6y^7; x, y \neq 0$

4) The result of the expression $(\frac{2}{5})^{-1} = \frac{5}{2} = 2.5$

5) The simplified expression of $(\frac{8x^{-5}}{24xy^{-3}})^{-1} = \frac{24xy^{-3}}{8x^{-5}} = 3\frac{x^6}{y^3} ; x, y \neq 0$

1.B. introduction to powers

b. Radicals, rules

1) The root of $\sqrt[4]{81x^4}$ is $\sqrt[4]{3^4x^4} = 3x$

2) The perimeter of a rectangle with length $L = 3\sqrt{20} + 5\sqrt{5}$ and width $W = 5\sqrt{20}$ is
$P = 2(L + W) = 2(3\sqrt{20} + 5\sqrt{5} + 5\sqrt{20}) = 2(3\sqrt{4*5} + 5\sqrt{5} + 5\sqrt{4*5})$
$$= 2(3\sqrt{4} * \sqrt{5} + 5\sqrt{5} + 5\sqrt{4} * \sqrt{5}) = 2(3 * 2\sqrt{5} + 5\sqrt{5} + 5 * 2\sqrt{5})$$
$$= 2(6\sqrt{5} + 5\sqrt{5} + 10\sqrt{5}) = 2(21\sqrt{5}) = 42\sqrt{5}$$

3) The simplified expression of $\frac{3\sqrt[3]{x^5y^3}}{27\sqrt{x^2y^4}}$ is $\frac{\sqrt[3]{x^3y^3x^2}}{9\sqrt{x^2}*\sqrt{y^4}} = \frac{xy\sqrt[3]{x^2}}{9xy^2} = \frac{\sqrt[3]{x^2}}{9y}$

4) The numerical value of $\sqrt[3]{-81} \div \sqrt[3]{-3}$ is $\frac{\sqrt[3]{-3^3*3}}{\sqrt[3]{-3}} = \frac{3\sqrt[3]{-3}}{\sqrt[3]{-3}} = 3$

5) The simplified form of $\frac{\sqrt{\sqrt{16}}}{\sqrt{\sqrt{625}}}$ is $\frac{\sqrt{\sqrt{16}}}{\sqrt{\sqrt{625}}} = \frac{\sqrt{4}}{\sqrt{25}} = \frac{2}{5}$

1.B. introduction to powers

c. Mixed radicals, conversions of radicals

1) The mixed radical of $\sqrt{192}$ is $\sqrt{64*3} = \sqrt{2^6*3} = 2^3\sqrt{3} = 8\sqrt{3}$

2) The mixed radical of $\sqrt{175}$ is $\sqrt{25*7} = \sqrt{25}*\sqrt{7} = 5\sqrt{7}$

3) The mixed radical of $\sqrt{288}$ is

$\sqrt{288} = \sqrt{2*144} = \sqrt{144}*\sqrt{2} = 12\sqrt{2}$

4) The entire radical of $8\sqrt{7}$ is

$8\sqrt{7} = \sqrt{64*7} = \sqrt{448}$

5) The entire radical of $6\sqrt{3}$ is

$6\sqrt{3} = \sqrt{36*3} = \sqrt{108}$

1.D. Mathematical expressions - Introduction

Substitute a value of n=3 into these expressions and find the value of R:

1) $R = 3n$ 2) $R = 5n - 3$ 3) $R = 2 \times (3n + 1)$

1) $R = 3(3) = 9$
2) $R = 5(3) - 3 = 15 - 3 = 12$
3) $R = 2 \times (3 \times 3 + 1) = 2 \times (9 + 1) = 2 \times 10 = 20$

Substitute a value of p=-4 into these expressions and find the value of R:

4) $R = (3p + 2)(p - 4)$
5) $R = 5 \times (p - 6) - 2p$
6) $R = 3 \times (-2p + 3)$

4) $R = [3(-4) + 2][(-4) - 4] = [-12 + 2](-8) = (-10)(-8) =$

5) $R = 5 \times (-4 - 6) - 2(-4) = 5 \times (-10) + 8 = -50 + 8 = -42$

6) $R = 3 \times [-2(-4) + 3] = 3 \times (8 + 3) = 3 \times 11 = 33$

1.E. Ordered pairs - Introduction

Represent the following points on the graph below.

A (1,1)	B (3,5)
C (-2,4)	D (-3,1)
E (-3,-5)	F (-4,0)
G (4,-2)	H (3,-3)

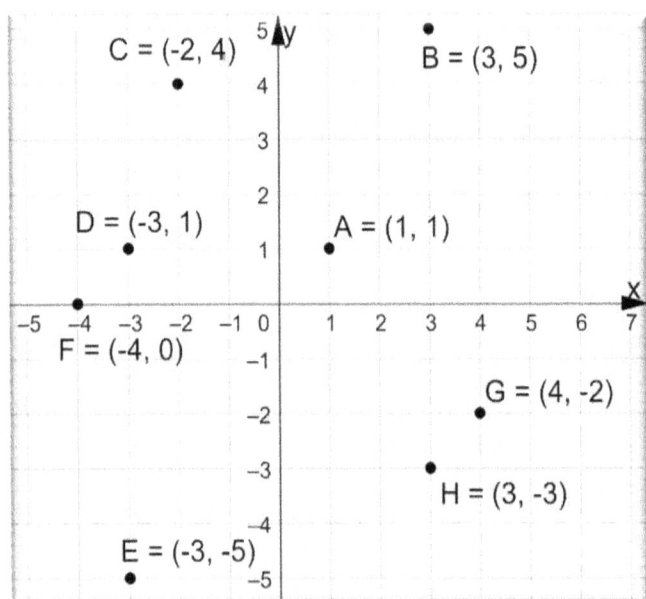

In what quadrants are situated the points below?

A (1,1)	Q1	B (3,5)	Q1
C (-2,4)	Q2	D (-3,1)	Q2
E (-3,-5)	Q3	F (-4,0)	Q2
G (4,-2)	Q4	H (3,-3)	Q4

CHAPTER 2

2.A. Representing patterns in linear relations

1) Analyze the pattern shown in the figures below. Find how many houses figure 5 will have.

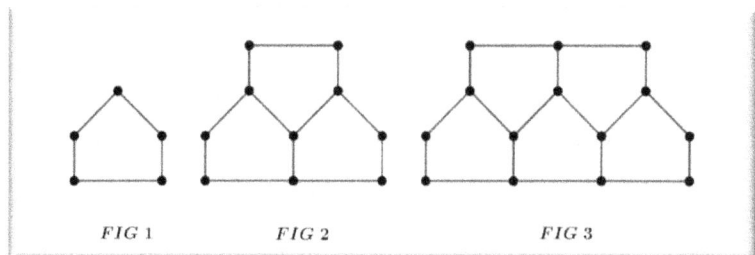

Figure	# Houses
1	1
2	3
3	5
5	7

2) Analyze the pattern shown in figures below. Find how many squares figure 5 will have.

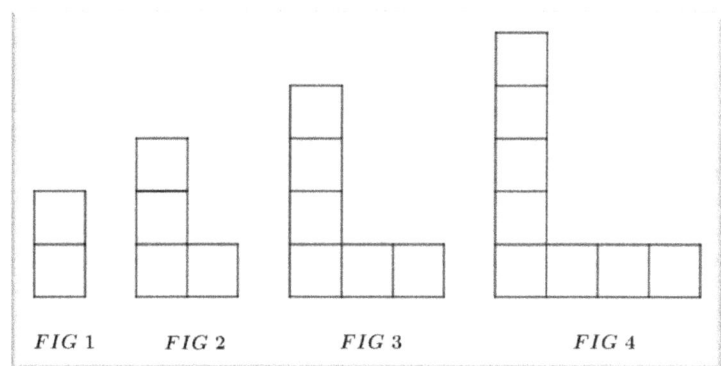

Figure	# Squares
1	2
2	4
3	6
4	8
5	10

3) Analyze the pattern shown in figures below. Find how many triangles figure 5 will have.

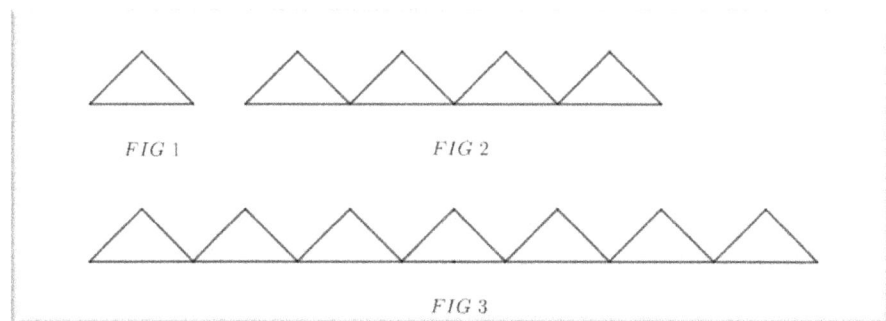

Figure	# Triangles
1	1
2	4
3	7
4	10
5	13

4) Victoria needs to build cardboard cubes. After building the fourth cube, she is left with 31 carboard pieces. Can she finish the tenth cube?

Cube	# Carboard pieces
1	6
2	12
3	18
4	24

For the ninth cube she will need 9*6=54 pieces

After the fourth cube she still has 31 pieces

54-24=30 pieces needed

She can build only nine cubes

For the tenth cube she will need 10*6=60 pieces

5) A rental car business has the client pay $40 for the first hour and $5 for every hour after that. How much does it cost to rent the car for 12 hours?

Cost=$40+$5*12h=$40+$60=$100

2.B. Understanding tables of values of linear relationships

1) In the relation $M = 3k - 4$ determine M when k is:

a) $M = 3k - 4 = 3(2) - 4 = 6 - 4 = 2$

b) $M = 3k - 4 = 3(17) - 4 = 51 - 4 = 47$

c) $M = 3k - 4 = 3(40) - 4 = 120 - 4 = 116$

d) $M = 3k - 4 = 3(y + 3) - 4 = 3y + 9 - 4 = 3y + 5$

2) Determine the common difference in the pattern below.

$$\sqrt{3} + \sqrt{5}, \quad \sqrt{3}, \quad \sqrt{3} - \sqrt{5} \ldots \ldots \ldots$$

$$\sqrt{3} - (\sqrt{3} + \sqrt{5}) = \sqrt{3} - \sqrt{3} - \sqrt{5} = -\sqrt{5}$$
$$\sqrt{3} - \sqrt{5} - \sqrt{3} = -\sqrt{5}$$

Common difference is $-\sqrt{5}$

3) Analyze the table below and write a relation between x and y.

X	1	2	3	4	5
Y	15	10	5	0	-5

Step 1

Check the difference between two consecutive values in the y column.

In this case that difference is -5

Step 2

Form the equation $y = -5 \times x$

Step 3

Check if for $x = 1$, the value for y at step 2 equals the value we should get in column y.

For $x = 1$, $y = -5 \times (1) = -5$

Step 4

We add or subtract any value from $A \times x$ in such a way that we obtain the value of y that is beside the value of x in the x column.

The value of y that corresponds to x=1 is 15 not -5.

We have to add 20 units to $y = -5 \times x$ in order to get 15.

The relation between y and x in this case will be:

$y = -5 \times x + 20$

Step 5

Check for another value of x if we obtain the correct corresponding value of y.

Let us take $x = 3$

Then we have: $y = -5 \times (3) + 20 = -15 + 20 = 5$ which equals the value of y for $x = 3$.

The equation that connects x and y is indeed $y = -5x + 20$

4) Determine the 20th term of the billow linear pattern

5,9,13,17,21

We create the table:

X (term #)	1	2	3	4	5
Y	5	9	13	17	21

Step 1

Check the difference between two consecutive values in the y column.

In this case that difference is 4

Step 2

Form the equation $y = 4 \times x$

Step 3

Check if for $x = 1$, the value for y at step 2 equals the value we should get in column y.

For $x = 1$, $y = 4 \times (1) = 4$

Step 4

We add or subtract any value from $A \times x$ in such a way that we obtain the value of y that is beside the value of x in the x column.

The value of y that corresponds to x=1 is 4 not 5.

We have to add 1 unit to $y = 4 \times x$ in order to get 5.

The relation between y and x in this case will be:

$y = 4 \times x + 1$

Step 5

Check for another value of x if we obtain the correct corresponding value of y.

Let us take $x = 3$

Then we have: $y = 4 \times (3) + 1 = 12 + 1 = 13$ which equals the value of y for $x = 3$.

The equation that connects x and y is indeed $y = 4x + 1$

The 50th term will be for x=50

$y(50) = 4(50) + 1 = 200 + 1 = 201$

5) The total cost for a publishing company to publish a book ,is a fixed cost (100) plus a cost for each additional book that the company will print. Create a general relation between the number of printed books and the cost of printing.

X (# of Books)	0	100	200	300	400
Cost	100	300	500	700	900

If we consider the situation for one book, two books and so on we will have the table:

X (# of Books)	0	1	2	3	4
Cost	100	3	5	7	9

Step 1

Check the difference between two consecutive values in the y column.

In this case that difference is 2

Step 2

Form the equation $y = 2 \times x$

Step 3

Check if for $x = 1$, the value for y at step 2 equals the value we should get in column y.

For $x = 1, y = 2 \times (1) = 2$

Step 4

We add or subtract any value from $A \times x$ in such a way that we obtain the value of y that is beside the value of x in the x column.

The value of y that corresponds to x=1 is 2 not 3.

We have to add 1 unit to $y = 2 \times x$ in order to get 3.

The relation between y and x in this case will be:

$y = 2 \times x + 1$

Step 5

Check for another value of x if we obtain the correct corresponding value of y.

Let us take $x = 3$

Then we have: $y = 2 \times (3) + 1 = 6 + 1 = 7$ which equals the value of y for $x = 3$.

The relation that connects the number of books printed and cost is indeed $y = 2x + 1$ plus the initial cost of $100

2.C. Understanding graphs of linear relationships

1) Represent the points from the table below and see if they are part of a straight line.

Point	X	Y
A	1	-4
B	2	0
C	3	4
D	4	8

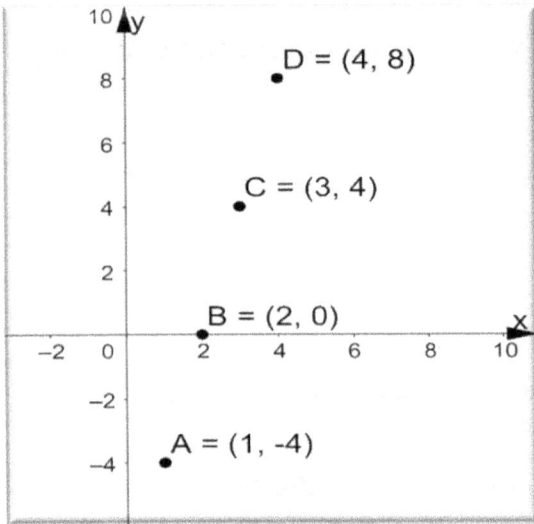

Part of a straight line

2) Represent graphically the points from the table below and see if they are part of a straight line.

Point	X	Y
A	1	-3
B	2	-1
C	3	1
D	4	3

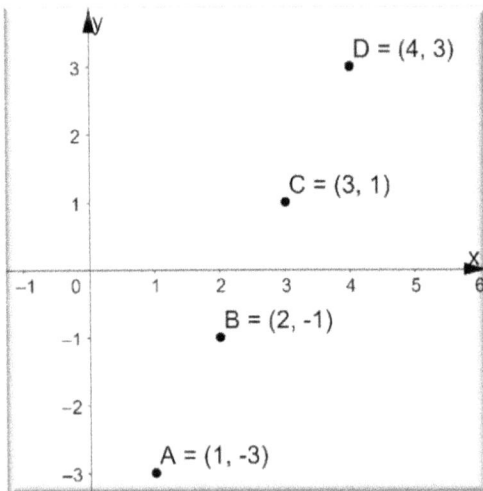

Part of a straight line

3) Represent graphically the points from the table below and see if they are part of a straight line.

Point	X	Y
A	1	2
B	2	4
C	4	5
D	5	8

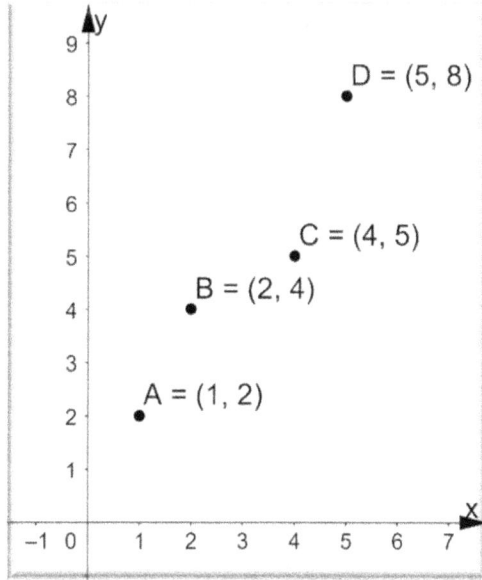

NOT Part of a straight line

4) Represent graphically the points from the table below and see if they are part of a straight line.

Point	X	Y
A	1	1
B	2	3
C	3	5

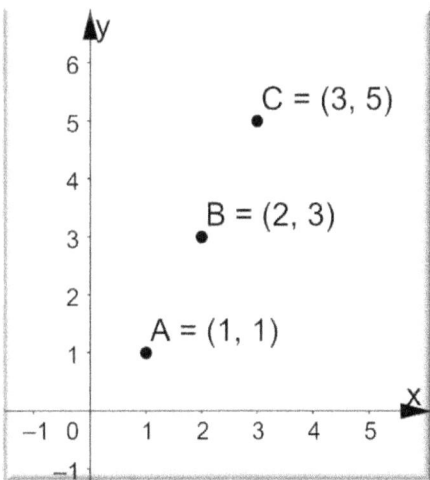

Part of a straight line

5) Represent graphically the points from the table below and see if they are part of a straight line.

Point	X	Y
A	1	2
B	3	-3
C	5	6
D	4	7

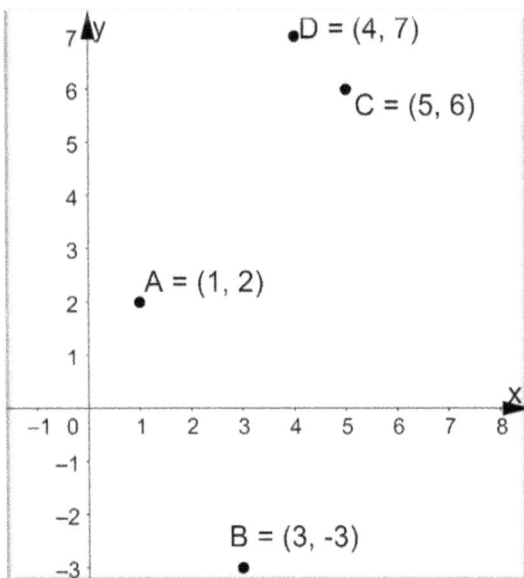

NOT Part of a straight line

2.D. Distance between points

a. Horizontal distance

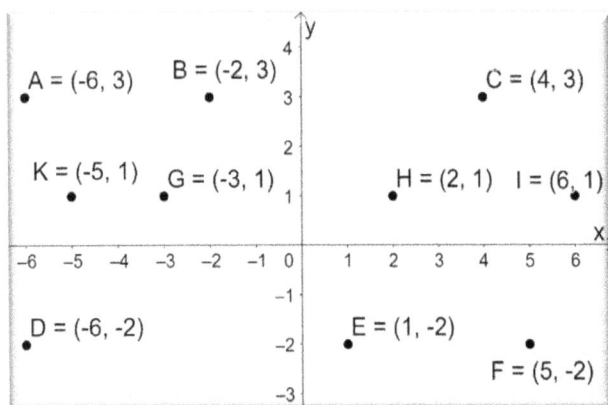

1) Incorrect

The distance between point A and point B is 5

$|-2-(-6)| = |4| = 4$

2) Incorrect

The distance between point E and point F is 7

$|5-(1.| = |4| = 4$

3) Correct

The distance between point K and point H is 7

$|2-(-5)| = |7| = 7$

4) Correct

The distance between point G and point H is 5

$|2 - (-3)| = |5| = 5$

5) Correct

The distance between point H and point I is 4

$|6 - 2| = |4| = 4$

2.D. Distance between points

b. Vertical distance

1) Incorrect

The distance between point A and point K is 10

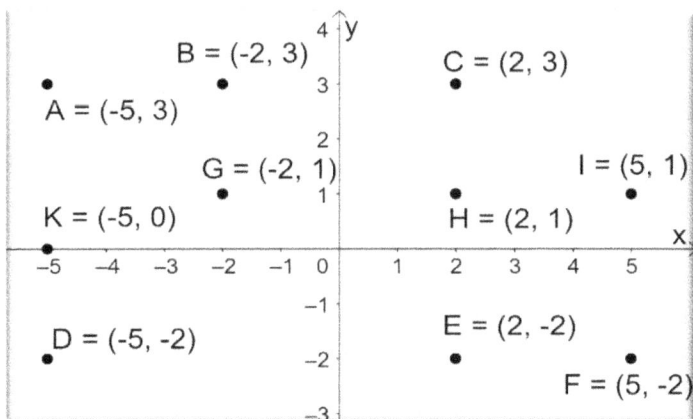

$|0 - 3| = |-3| = 3$

2) Correct

The distance between point A and point D is 5

$|-2 - 3| = |-5| = 5$

3) Incorrect

The distance between point B and point G is 3

$|1 - 3| = |-2| = 2$

4) Incorrect

The distance between point C and point H is 7

$|1 - 3| = |-2| = 2$

5) Correct

The distance between point C and point E is 5

$|-2 - 3| = |-5| = 5$

6) Incorrect

The distance between point K and point D is 7

$|-2 - (0)| = |-2| = 2$

7) Correct

The distance between point H and point E is 3

$|-2 - 1| = |-3| = 3$

8) Incorrect

The distance between point I and point F is 7

$|-2 - (-6)| = |4| = 4$

9) Incorrect

The distance between point A and point K is 9

$|0 - 3| = |-3| = 3$

10) Incorrect

The distance between point I and point H is 4

$|5 - 2| = |3| = 3$ (*Horizontal distance*)

2.D. Distance between points

c. Non-horizontal and Non-vertical distance

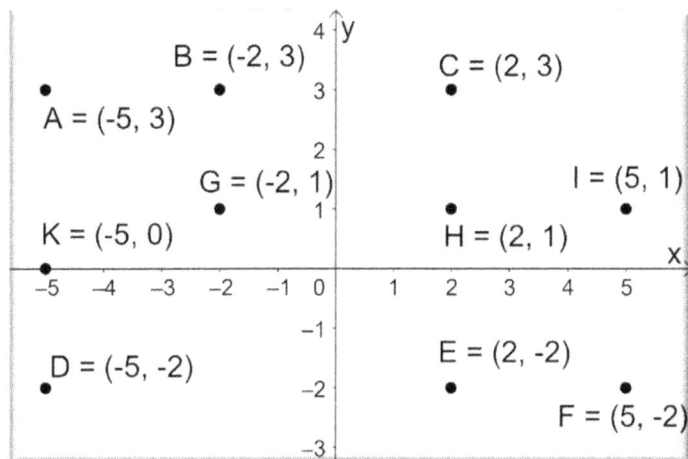

B = (-2, 3) C = (2, 3)

A = (-5, 3)

G = (-2, 1) I = (5, 1)

K = (-5, 0) H = (2, 1)

D = (-5, -2) E = (2, -2)

F = (5, -2)

1) Correct

The distance between point B and point H is $2\sqrt{5} = 4.47$

$BH=\sqrt{(x_2 - x_1)^2 + (y_2 - y_1)^2} = \sqrt{[(2 - (-2)]^2 + (1 - 3)^2} = \sqrt{16 + 4} = \sqrt{20} = \sqrt{4 \times 5} = 2\sqrt{5}$

2) Incorrect

The distance between point K and point C is 7.61

$GC=\sqrt{(x_2 - x_1)^2 + (y_2 - y_1)^2} = \sqrt{(2 - (-5))^2 + (3 - 0)^2} = \sqrt{49 + 9} = \sqrt{58} = 7.61$

3) Correct

The distance between point A and point H is

$AH=\sqrt{(x_2 - x_1)^2 + (y_2 - y_1)^2} = \sqrt{(2 - (-5))^2 + (1 - 3)^2} = \sqrt{49 + 4} = \sqrt{53} = 7.28$

4) Correct

The distance between point A and point F is 11.18

$AF=\sqrt{(x_2 - x_1)^2 + (y_2 - y_1)^2} = \sqrt{(5 - (-5))^2 + (-2 - 3)^2} = \sqrt{100 + 25} = \sqrt{125} = 11.18$

5) Incorrect

The distance between point B and point D is 5.83

$BD=\sqrt{(x_2 - x_1)^2 + (y_2 - y_1)^2} = \sqrt{(-5 - (-2))^2 + (-2 - 3)^2} = \sqrt{9 + 25} = \sqrt{34} = 5.83$

6) Incorrect

The distance between point B and point F is 8.6

$BF=\sqrt{(x_2 - x_1)^2 + (y_2 - y_1)^2} = \sqrt{(5 - (-2))^2 + (-2 - 3)^2} = \sqrt{49 + 25} = \sqrt{74} = 8.6$

2.D. Distance between points

d. Midpoint coordinates

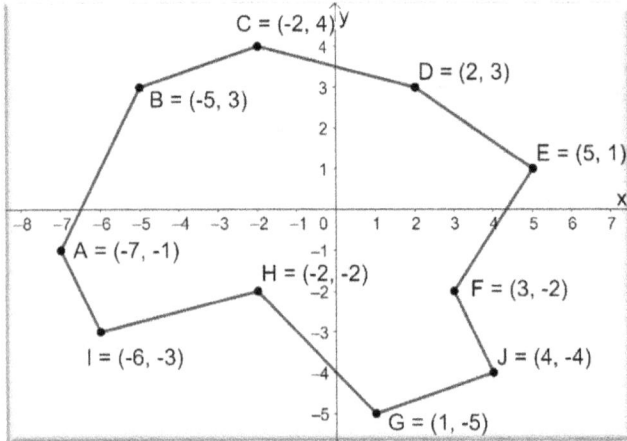

1) Correct

The mid-point coordinates of segment AB are:

$$x = \frac{x_2 + x_1}{2} = \frac{-5 + (-7)}{2} = \frac{-12}{2} = -6$$

$$y = \frac{y_2 + y_1}{2} = \frac{3 + (-1.}{2} = \frac{2}{2} = 1$$

2) Incorrect

The mid-point coordinates of segment BC are:

$$x = \frac{x_2 + x_1}{2} = \frac{-2 + (-5)}{2} = \frac{-7}{2} = -3.5$$

$$y = \frac{y_2 + y_1}{2} = \frac{4 + 3}{2} = \frac{7}{2} = 3.5$$

3) Correct

The mid-point coordinates of segment CD are:

$$x = \frac{x_2 + x_1}{2} = \frac{2 + (-2)}{2} = \frac{0}{2} = 0$$

$$y = \frac{y_2 + y_1}{2} = \frac{4 + 3}{2} = \frac{7}{2} = 3.5$$

4) Incorrect

The mid-point coordinates of segment DE are:

$$x = \frac{x_2 + x_1}{2} = \frac{2 + 5}{2} = \frac{7}{2} = 3.5$$

$$y = \frac{y_2 + y_1}{2} = \frac{1 + 3}{2} = \frac{4}{2} = 2$$

5) Incorrect

The mid-point coordinates of segment EF are:

$$x = \frac{x_2 + x_1}{2} = \frac{3 + 5}{2} = \frac{8}{2} = 4$$

$$y = \frac{y_2 + y_1}{2} = \frac{-2 + 1}{2} = \frac{-1}{2} = -0.5$$

6) Correct

The mid-point coordinates of segment FJ are:

$$x = \frac{x_2 + x_1}{2} = \frac{3 + 4}{2} = \frac{7}{2} = 3.5$$

$$y = \frac{y_2 + y_1}{2} = \frac{-2 - 4}{2} = \frac{-6}{2} = -3$$

7) Correct

The mid-point coordinates of segment JG are:

$$x = \frac{x_2 + x_1}{2} = \frac{1+4}{2} = \frac{5}{2} = 2.5$$

$$y = \frac{y_2 + y_1}{2} = \frac{-5+(-4)}{2} = \frac{-9}{2} = -4.5$$

8) Incorrect

The mid-point coordinates of segment GH are:

$$x = \frac{x_2 + x_1}{2} = \frac{-2+1}{2} = \frac{-1}{2} = -0.5$$

$$y = \frac{y_2 + y_1}{2} = \frac{-2+(-5)}{2} = \frac{-7}{2} = -3.5$$

9) Correct

The mid-point coordinates of segment HI are:

$$x = \frac{x_2 + x_1}{2} = \frac{-6+(-2)}{2} = \frac{-8}{2} = -4$$

$$y = \frac{y_2 + y_1}{2} = \frac{-3+(-2)}{2} = \frac{-5}{2} = -2.5$$

10) Incorrect

The mid-point coordinates of segment IA are:

$$x = \frac{x_2 + x_1}{2} = \frac{-7+(-6)}{2} = \frac{-13}{2} = -6.5 \quad y = \frac{y_2 + y_1}{2} = \frac{-1+(-3)}{2} = \frac{-4}{2} = -2$$

2.E. Slope of a line

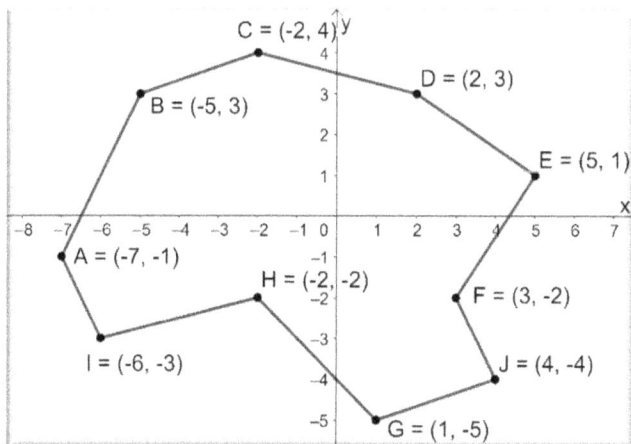

1) Correct

The slope of segment AB is:

$$m = \frac{y_2 - y_1}{x_2 - x_1} = \frac{3-(-1.}{-5-(-7)} = \frac{4}{2} = 2$$

2) Incorrect

The slope of segment BC is

$$m = \frac{y_2 - y_1}{x_2 - x_1} = \frac{4-3}{-2-(-5)} = \frac{1}{3}$$

3) Correct

The slope of segment CD is

$$m = \frac{y_2 - y_1}{x_2 - x_1} = \frac{3-4}{2-(-2)} = \frac{-1}{4}$$

4) Incorrect

The slope of segment DE is

$$m = \frac{y_2 - y_1}{x_2 - x_1} = \frac{1-3}{5-2} = \frac{-2}{3}$$

5) Incorrect

The slope of segment EF is

$$m = \frac{y_2 - y_1}{x_2 - x_1} = \frac{-2-1}{3-5} = \frac{-3}{-2} = 1.5$$

6) Correct

The slope of segment FJ is:

$$m = \frac{y_2 - y_1}{x_2 - x_1} = \frac{-4-(-2)}{4-3} = \frac{-2}{1} = -2$$

7) Correct

The slope of segment JG is

$$m = \frac{y_2 - y_1}{x_2 - x_1} = \frac{-5-(-4)}{1-4} = \frac{-1}{-3} = \frac{1}{3}$$

8) Incorrect

The slope of segment GH is

$$m = \frac{y_2 - y_1}{x_2 - x_1} = \frac{-2-(-5)}{-2-1} = \frac{3}{-3} = -1$$

9) Correct

The slope of segment HI is

$$m = \frac{y_2 - y_1}{x_2 - x_1} = \frac{-3-(-2)}{-6-(-2)} = \frac{-1}{-4} = \frac{1}{4}$$

10) Incorrect

The slope of segment IA is

$$m = \frac{y_2 - y_1}{x_2 - x_1} = \frac{-1-(-3)}{-7-(-6)} = \frac{2}{-1} = -2$$

CHAPTER 3

3.B. Solving Linear Equations

a. Solve one-step linear equations:

1) Solve ad check the solution.

a) $x + 3 = 25$ b) $x + \frac{1}{2} = 3\frac{1}{3}$

a) $x + 3 - 3 = 25 - 3$ We subtract 3 on each side

$x + 0 = 22$

Check

$22 + 3 = 25$

$25 = 25$

b) $x + \frac{1}{2} = 3\frac{1}{3}$ We subtract $\frac{1}{2}$ from both sides of the equation

$x + \frac{1}{2} - \frac{1}{2} = 3\frac{1}{3} - \frac{1}{2}$

$x + 0 = 3\frac{1}{3} - \frac{1}{2}$ We transform the mixed fraction into an improper fraction

$x = \frac{10}{3} - \frac{1}{2}$ We find the common denominator as 6

$x = \frac{10 \times 2}{3 \times 2} - \frac{1 \times 3}{2 \times 3}$

$x = \frac{20}{6} - \frac{3}{6}$

$x = \frac{20 - 3}{6} = \frac{17}{6} = 2\frac{5}{6}$

Check

$\frac{17}{6} + \frac{1}{2} = 3\frac{1}{3}$ We find that the common denominator is 6

$\frac{17}{6} + \frac{1 \times 3}{2 \times 3} = 3\frac{1}{3}$

$\frac{17}{6} + \frac{3}{6} = 3\frac{1}{3}$

$\frac{17 + 3}{6} = 3\frac{1}{3}$

$\frac{20}{6} = 3\frac{1}{3}$

$3\frac{2}{6} = 3\frac{1}{3}$ We divide fraction $\frac{2}{6}$ with 2 at the denominator and numerator

$3\frac{1}{3} = 3\frac{1}{3}$

2) Write each sentence as an equation

a) x decreased by 3 is 7 b) The sum of 4 and x is 9

$x - 3 = 7$ $x + 4 = 9$

3) Check if the given value of x is a solution for the equation.

a) $x + 3 = 7$ for $x = 5$

b) $-x - 2 = 7$ for $x = 4$

c) $x + 5 = 7$ for $x = 2$

d) $x + 3.4 = 5.8$ for $x = 4.3$

a) $5 + 3 = 7$

$8 \neq 7$ $x = 5$ is not a solution

b) $-4 - 2 = 7$

$-6 \neq 7$ $x = 4$ is not a solution

c) $2 + 5 = 7$

$7 = 7$ $x = 2$ is the solution

d) $4.3 + 3.4 = 5.8$

$7.7 \neq 5.8$ $x = 4.3$ is not a solution

4) Explain the error.

$x - \dfrac{2}{7} = \dfrac{5}{7}$

$x = \dfrac{3}{7}$

Error: we have to add $\dfrac{2}{7}$ on both sides. So, we will have $x = \dfrac{2}{7} + \dfrac{5}{7} = \dfrac{7}{7} = 1$

5) The 30cm Length of a rectangle is bigger that the Width by 12cm. How much is the Perimeter?

So,

$30 - 12 = W$

$W = 18cm$

Perimeter $= 2(30) + 2(18) = 60 + 36 = 96cm$

3.B. Solving Linear Equations

b. Solving two-step linear equations with addition and subtraction

1) Solve and check

$2x - 4 = x + 2$

$2x - 4 = x + 2$ We add 4 on each side

$2x - 4 + 4 = x + 2 + 4$

$2x - 0 = x + 6$ We subtract x on each side

$2x - x = x - x + 6$

$x = 0 + 6$

$x = 6$

Check

$2(6) - 4 = 6 + 2$

$12 - 4 = 8$

$8 = 8$

$x = 6$ is the solution of the equation $2x - 4 = x + 2$

2) Solve and check.

$5x + \frac{1}{3} = 4x + 1\frac{1}{2}$

$5x + \frac{1}{3} = 4x + 1\frac{1}{2}$ We subtract $\frac{1}{3}$ on each side

$5x + \frac{1}{3} - \frac{1}{3} = 4x + 1\frac{1}{2} - \frac{1}{3}$

$5x + 0 = 4x + 1\frac{1}{2} - \frac{1}{3}$ We subtract $4x$ on each side

$5x - 4x = 4x - 4x + 1\frac{1}{2} - \frac{1}{3}$

$x = 0 + 1\frac{1}{2} - \frac{1}{3}$ We transform $1\frac{1}{2}$ into an improper fraction

$x = \frac{3}{2} - \frac{1}{3}$ We find common denominator being 6

$x = \frac{3\times3}{2\times3} - \frac{1\times2}{3\times2}$

$x = \frac{9}{6} - \frac{2}{6} = \frac{9-2}{6} = \frac{7}{6} = 1\frac{1}{6}$

$x = 1\frac{1}{6}$

Check

$5(\frac{7}{6}) + \frac{1}{3} = 4(\frac{7}{6}) + 1\frac{1}{2}$ We transform $1\frac{1}{2}$ into an improper fraction

$5(\frac{7}{6}) + \frac{1}{3} = 4(\frac{7}{6}) + \frac{3}{2}$ We multiply 5 and 4 with $\frac{7}{6}$

$\frac{35}{6} + \frac{1}{3} = \frac{28}{6} + \frac{3}{2}$ We multiply each term with 6 The common

denominator

$\frac{35\times6}{6} + \frac{1\times6}{3} = \frac{28\times6}{6} + \frac{3\times6}{2}$ We simplify wherever possible.

$35 + 2 = 28 + 9$

$37 = 37$

$x = 1\frac{1}{6}$ is the solution of the equation $5x + \frac{1}{3} = 4x + 1\frac{1}{2}$

3) Solve ad check

$3.25x - 5 = 2.25x - 6$

$3.25x - 5 = 2.25x - 6$ We add 5 on each side

$3.25x - 5 + 5 = 2.25x - 6 + 5$

$3.25x + 0 = 2.25x - 1$ We subtract $2.25\,x$ on each side

$3.25x - 2.25x = 2.25x - 2.25x - 1$

$x = -1$

Check

$3.25(-1) - 5 = 2.25(-1) - 6$

$3.25(-1) - 5 = 2.25(-1) - 6$

$-3.25 - 5 = -2.25 - 6$

$-8.25 = -8.25$

$x = -1$ is the solution of the equation $3.25x - 5 = 2.25x - 6$

4) Check if these given value of x are the solution for the equation below.

$3x - 5 = 5x + 3$ $x = 2, -3, -4$

For $x = 2$

$3(2) - 5 = 5(2) + 3$

$6 - 5 = 10 + 3$

$1 \neq 13$ $x = 2$ is not a solution of the equation $3x - 5 = 5x + 3$

For $x = -3$

$3(-3) - 5 = 5(-3) + 3$

$-9 - 5 = -15 + 3$

$-14 \neq -12$

$x = -3$ is not a solution for the equation $3x - 5 = 5x + 3$

For $x = -4$

$3(-4) - 5 = 5(-4) + 3$

$-12 - 5 = -20 + 3$

$-17 = -17$

$x = -4$ is a solution for the equation $3x - 5 = 5x + 3$

5) Write an equation. Solve and check.

Four times a number increased by 3 is 3 times a number decreased by 1

$4x + 3 = 3x - 1$ We subtract 3 on both sides.

$4x + 3 - 3 = 3x - 1 - 3$

$4x = 3x - 4$ We subtract 3x on each side

$4x - 3x = 3x - 3x - 4$

$x = -4$

Check

$4(-4) + 3 = 3(-4) - 1$

$-16 + 3 = -12 - 1$

$-13 = -13$

$x = -4$ is the solution of the equation $4x + 3 = 3x - 1$

3.B. Solving Linear Equations

c. Solving two-step linear equations with multiplication and division

1) Solve: $5x - 2 = 4$

In this case, to isolate the variable x, we have to add 2 and then divide both sides of the equation with the same value 5.

$5x - 2 + 2 = 4 + 2$

$5x = 6$

$\frac{5}{5}x = \frac{6}{5}$

$x = \frac{6}{5}$

2) Solve and check $4x + 3 = 4 + x$

We will have to ISOLATE the unknown or the variable.

We will do the same operations in both terms so the equation is balanced all the time.

Step 1: minus 3 in both sides of the equation.

$4x + 3 - 3 = 4 - 3 + x$

$4x = 1 + x$

Step 2: subtract x in each side.

$4x - x = 1 + x - x$

$3x = 1$

Step 3: divide by 3 in both sides

$\frac{3x}{3} = \frac{1}{3}$

$x = \frac{1}{3}$

To be sure the result is correct, we check by substituting $x = \frac{3}{2}$ in the original equation.

Check

$4\left(\frac{1}{3}\right) + 3 = 4 + \frac{1}{3}$

$\frac{4}{3} + \frac{9}{3} = \frac{12}{3} + \frac{1}{3}$

$\frac{13}{3} = \frac{13}{3}$

Indeed, $x = \frac{1}{3}$ is the <u>solution</u> of the equation $4x + 3 = 4 + x$

3) Solve $\frac{5}{x} = 3$ $x \neq 0$

In this case, to isolate the variable x, we have to:

Step 1 multiply by x both sides of the equation

$\frac{5}{x} * x = 3x$

$5 = 3x$

Step 2 divide both sides of the equation with the same value 3.

Here we divide by 4

$\frac{5}{3} = \frac{3x}{3}$

$x = \frac{5}{3}$

4) Solve and check $3.2x + 5.2 = 9.4 - 2.1x$

We will have to ISOLATE the unknown or the variable.

We will do the same operations in both terms so the equation is balanced all the time.

Step 1: minus 5.2 in both sides of the equation.

$3.2x + 5.2 - 5.2 = 9.4 - 5.2 - 2.1x$

$3.2x = 4.2 - 2.1x$

Step 2: add $2.1x$ in both sides

$3.2x + 2.1x = 4.2 - 2.1x + 2.1x$

$5.3x = 4.2$

Step 3 divide by 5.3 in each side of the equation.

$\dfrac{5.3x}{5.3} = \dfrac{4.2}{5.3}$

$x = 0.79$

To be sure the result is correct, we check by substituting $x = 0.79$ in the original equation.

Check

$3.2(0.79) + 5.2 = 9.4 - 2.1(0.79)$

$2.54 + 5.2 = 7.74$

$7.74 = 7.74$

Indeed, $x = 0.79$ is the <u>solution</u> of the equation $3.2x + 5.2 = 9.4 - 2.1x$

5) Fifteen divided by a number is 5. Write then solve an equation to determine the number. Verify the solution.

$\dfrac{15}{x} = 5$

In this case, to isolate the variable x, we have to multiply by x and then divide both sides of the equation with the same value 5.

$\dfrac{15x}{x} = 5x$

$15 = 5x$

$\dfrac{15}{5} = \dfrac{5x}{5}$

$x = 3$

Check

$\dfrac{15}{3} = 5$

$5 = 5$

$x = 3$ is the solution of the equation $\frac{15}{x} = 5$

3.B. Solving Linear Equations

d. Solving two-step linear equations with distributivity property

1) Solve and check

$3(x + 2) = 7$

Step 1 we multiply 3 with x and 2 respectively

$3x + 6 = 7$

Step 2 we subtract 6 in both sides

$3x + 6 - 6 = 7 - 6$

$3x = 1$

Step 3 we divide by 3 in both sides

$\frac{3x}{3} = \frac{1}{3}$

$x = \frac{1}{3}$

Check

$3\left(\frac{1}{3} + 2\right) = 7$

$3 * \left(\frac{7}{3}\right) = 7$

$7 = 7$

$x = \frac{1}{3}$ is the solution of the equation $3(x + 2) = 7$

2) Solve and check

$4(3x - 1) = 3(x + 5)$

Step 1 we multiply by 4 the first bracket and y 3 the second bracket

$12x - 4 = 3x + 15$

Step 2 we add 4 in both sides

$12x - 4 + 4 = 3x + 15 + 4$

$12x = 3x + 19$

Step 3 we subtract $3x$ in both sides

$12x - 3x = 3x - 3x + 19$

$9x = 19$

Step 4 we divide by 9 in both sides of the equation

$$\frac{9x}{9} = \frac{19}{9}$$

$x = 2.11$

Check

$4(3 * 2.11 - 1) = 3(2.11 + 5)$

$4(6.33 - 1) = 3(7.11)$

$4(5.33) = 21.33$

$21.33 = 21.33$

$x = 2.11$ is the solution of the equation $4(3x - 1) = 3(x + 5)$

3) Solve and check

$\frac{1}{2}(3x - 4) = \frac{3}{2}(2x + 5)$

Step 1 we multiply by 2 in both sides to get rid of the fractions

$\frac{2}{2}(3x - 4) = \frac{3*2}{2}(2x + 5)$

$3x - 4 = 3(2x + 5)$

Step 2 we multiply 3 with $2x$ and 5 respectively in the right side of the equation.

$3x - 4 = 6x + 15$

Step 3 we add 4 in both sides of the equation

$3x - 4 + 4 = 6x + 15 + 4$

$3x = 6x + 19$

Step 4 we subtract $6x$ in both sides

$3x - 6x = 6x - 6x + 19$

$-3x = 19$

Step 5 we divide by minus 3 each side

$-\frac{3x}{-3} = \frac{19}{-3}$

$x = -6.33$

Check

$\frac{1}{2}[3(-6.33) - 4] = \frac{3}{2}[2(-6.33) + 5]$

$\frac{1}{2}(-18.99 - 4) = \frac{3}{2}(-12.66 + 5)$

$\frac{1}{2}(-22.99) = 1.5(-7.66)$

$-11.49 = -11.49$

$x = -6.33$ is the solution for the equation $\frac{1}{2}(3x - 4) = \frac{3}{2}(2x + 5)$

4) Solve

$$\frac{x}{3} + \frac{5}{3} = \frac{3}{4}$$

Step 1 we multiply by the common denominator 12 in each side

$$\frac{12x}{3} + \frac{12*5}{3} = \frac{12*3}{4}$$

$$4x + 20 = 9$$

Step 2 we subtract 20 in each side of the equation

$$4x + 20 - 20 = 9 - 20$$

$$4x = -11$$

Step 3 we divide by 4 in each side

$$\frac{4x}{4} = -\frac{11}{4}$$

$$x = -2\frac{3}{4}$$

5) Solve

$$\frac{1}{3}(2x - 3) + 4x - 3 = \frac{5}{6}(x + 1) + 2$$

Step 1 We multiply by 6 all terms in each side of the equation

$$\frac{1*6}{3}(2x - 3) + 4*6x - 3*6 = \frac{5*6}{6}(x + 1) + 2*6$$

$$2(2x - 3) + 24 - 18 = 5(x + 1) + 12$$

Step 2 we apply the distributive property for each bracket.

$$4x - 6 + 6 = 5x + 5 + 12$$

$$4x = 5x + 17$$

Step 3 we subtract 5x in each side of the equation

$$4x - 5x = 5x - 5x + 17$$

$$-x = 17$$

Step 4 we multiply by −1 in each side

$$-x * (-1) = 17 * (-1)$$

$$x = -17$$

3.C. Equation of a straight line

a. Non-vertical and non-horizontal line

1) The equation of the line through M (-3,1) and slope -2 is $y = -2x - 5$

$$m = \frac{y-y_1}{x-x_1}$$

$$-2 = \frac{y-1}{x-(-3)}$$

$$-2x - 6 = y - 1$$

$$y = -2x - 5$$

2) The y intercept of the line $y = -2x - 5$ is y=-5

3) In the slope relation, $m = \frac{y-5}{x+4}$, the y intercept in terms of the slope m, is $b = 4m + 5$

$$m = \frac{y-5}{x+4}$$

$$m(x + 4) = y - 5$$

$$mx + 4m = y - 5$$

$$mx + 5 + 4m = y$$

$$y = mx + 4m + 5$$

$$y = mx + b$$

$$b = 4m + 5$$

4) The equation of the parallel line with $y = 3x + 1$ that passes through the point M (5,6) is $y = 3x - 9$

$$m = \frac{y-6}{x-5} = 3$$

$$3(x - 5) = y - 6$$

$$3x - 15 = y - 6$$

$$y = 3x - 9$$

5) The intersection to x axis of $y = 4x - 8 \ is$ P (2,0)

$$y = 0$$

So,

$$0 = 4x - 8$$

$$8 = 4x$$

$$x = \frac{8}{4} = 2$$

3.D. Straight-line graph

1) The slope of the line $y = 3x - 1$ is 3

2) The slope of the line $-3(x + 2) - 4(y - 7) = 6$ is $m = -\frac{3}{4}$

$-3(x + 2) - 4(y - 7) = 6$

$-3x - 6 - 4y + 28 = 6$

$-3x - 4y + 22 = 6$

$-4y = 3x - 16$

$4y = -3x + 16$

$y = -\frac{3}{4}x + 4$

The next problems use the figure shown below

3)

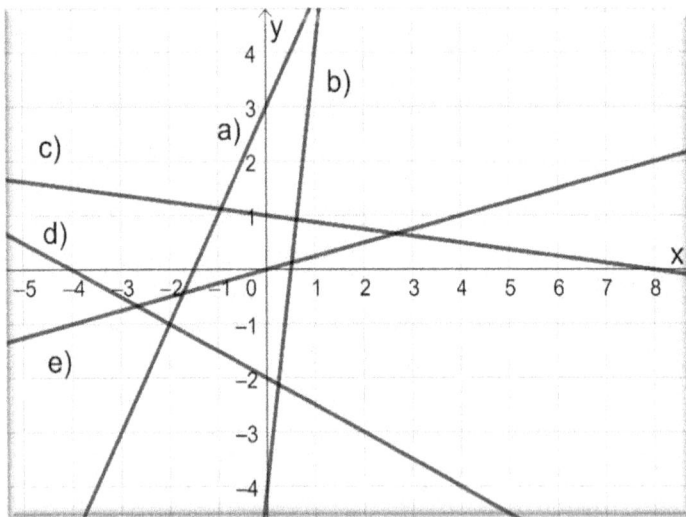

The line a) has the equation

Y=2x+3

4)

The line c) has the equation

$Y = -\frac{1}{8}x + 1$

5)

The line e) has the equation

$y = \frac{1}{4}x$

3.E. Special cases of linear equations:

a. Vertical and horizontal lines

1) The equation of horizontal line through M (3,4) is $y = 4$

Problems 2,3,4,5 will be based on the figure shown below.

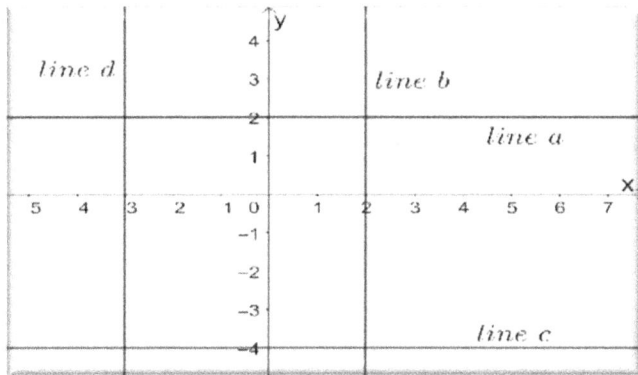

2) The equation of line a is $y = 2$

3) The equation of line b is $x = 3$

4) The equation of line c is $y = -4$

5) The equation of line d is $x = -3$

3.F. Parallel and perpendicular lines

a. Parallel lines

1) The equation of the parallel line with $y = x - 1$ that intersects y axis at point M (0,5) is $y = x + 5$

2) The line $y = -5x + 3$ is not parallel with $y = -4x + 3$

3) The line $y = 4x + 3$ is parallel with $y = 4x - 25$

Problems 4 and 5 will be based on the figure shown below.

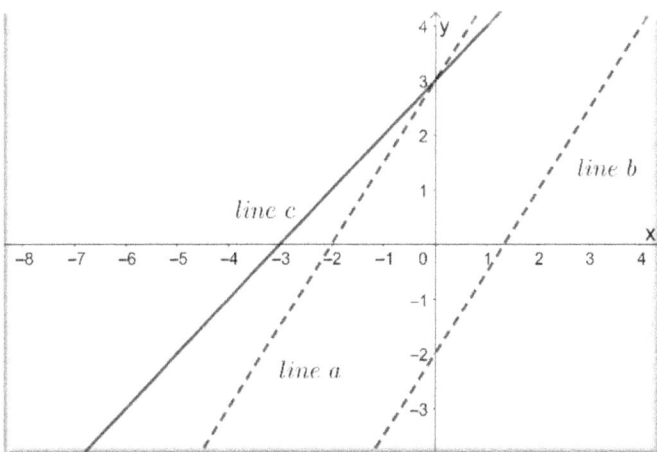

4) The line a is parallel to line b. $m_a = m_b = \frac{3}{2}$

5) The equation of the line b is $y = \frac{3}{2}x - 2$

3.F Parallel and perpendicular lines

b. Perpendicular lines

1) The lines that have the equations $y = 3x + 4$ and $y = -\frac{1}{3}x - 5$ are perpendicular.

2) The line perpendicular to the line $y = -2x + 7$ has the slope $m = \frac{1}{2}$.

3) The equation of the line perpendicular to $y = -2x + 3$ through M (-2,-3) is

$y = \frac{1}{2}x - 2$

The slope of the perpendicular line is $m = \frac{1}{2}$

We substitute the coordinates of M into the formula $y = mx + b$

$-3 = \frac{1}{2}(-2) + b$

$-3 = -1 + b$

$-3 + 1 = +b$

$b = -2$

So, the equation of the line perpendicular to $y = -2x + 3$ through M (-2,-3) is

$y = \frac{1}{2}x - 2$

4) The equation of the line perpendicular to AB in point B, is $y = \frac{7}{3}x - 11.33$

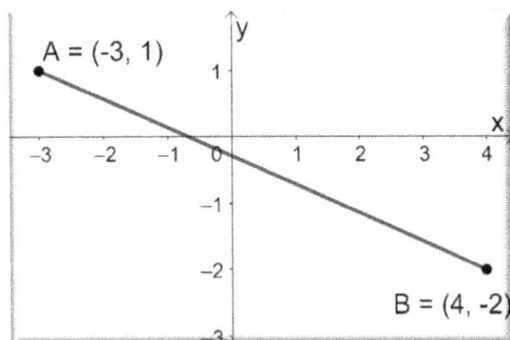

Slope of AB is $m = \frac{y_2 - y_1}{x_2 - x_1} = \frac{-2 - 1}{4 - (-3)} = \frac{-3}{7}$

So,

The slope of the perpendicular to AB is $m = \frac{7}{3}$

The perpendicular goes through B so, we substitute point B coordinates into $y = mx + b$

$-2 = \frac{7}{3}(4) + b$

$-2 - \frac{28}{3} = b$

$b = -11.33$

So, the equation of the line perpendicular to AB through B (4,-2) is $y = \frac{7}{3}x - 11.33$

5) The perpendicular to AB through the point B (6,1) will intersect y axis in $b = 13$

The slope of AB is $m = \frac{y_2-y_1}{x_2-x_1} = \frac{1-(-3)}{6-(-2)} = \frac{4}{8} = \frac{1}{2}$

So, the slope of the perpendicular to AB will be $m = -2$

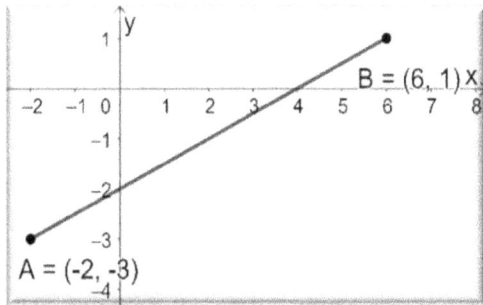

The perpendicular goes through B(6,1) so, we substitute point B coordinates into $y = mx + b$

$1 = -2(6) + b$

$1 + 12 = b$

$b = 13$

CHAPTER 4

4.A. Express linear inequalities graphically and algebraically

1) Represent on the number line and algebraically:

A number bigger than and equal to -5

$x \geq -5$

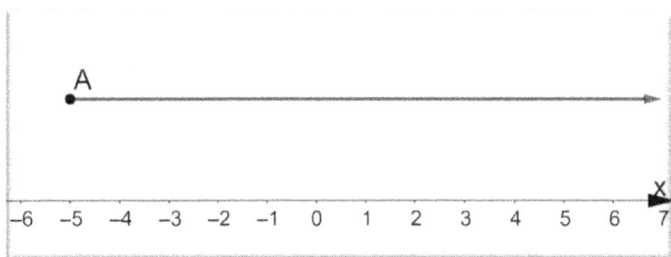

2) Represent on the number line and algebraically:

A number less than and equal to 2

$x \leq 2$

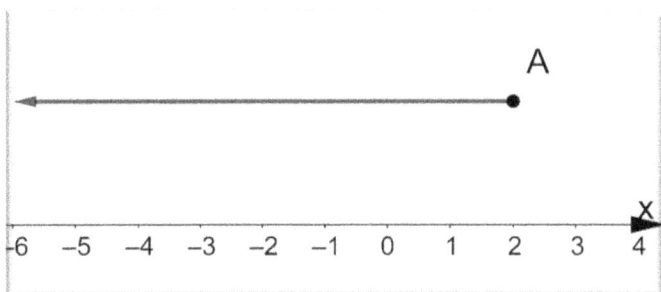

3) Use a symbol $>, <, \leq, or \geq$ to write an inequality that corresponds to each statement.

a) x is less than -7

$x < -7$

b) a number is greater and equal than 3

$x \geq 3$

c) x is negative

$x < 0$

4) Write 3 numbers that are solutions of each inequality.

a) $a > 3$ b) $b \leq 3$ c) $w < -5$

a) $c > 3$ b) $d \leq 3$ c) $w < -5$

4,5,6   -3, 0, 3   -10, -7.5, -4.99

5) Iveta and Emma write the inequality whose solution is shown below.

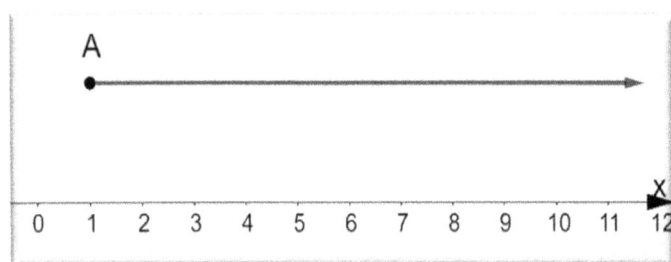

Iveta writes $x \geq 1$

Emma writes $1 > x$

Who is correct?

The solution are all the points that are bigger and equal with 1.

Iveta is correct, Emma is wrong (she wrote $x < 1$

4.B. Solving one-step linear inequalities

1) Solve and graph the solution on the number line.

$x - 7 \geq -4$

We add 7 in each side.

$x - 7 + 7 \geq -4 + 7$

$x \geq 3$

A

$$\begin{array}{ccccccccccc} | & | & | & \bullet & | & | & | & | & | & | & X \\ 0 & 1 & 2 & 3 & 4 & 5 & 6 & 7 & 8 & 9 \end{array}$$

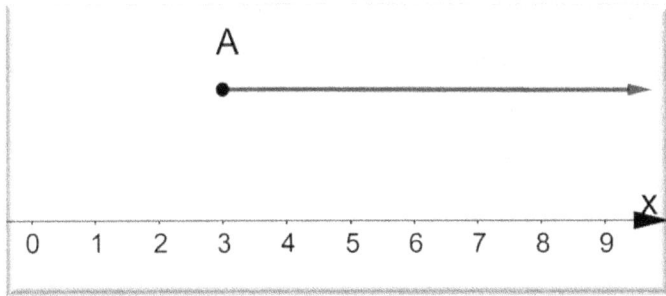

2) What must be done to the first inequality to get to the second inequality?

a) $x - 5 \leq 4$

$x \leq 9$

b) $2x \leq 4$

$x \leq 2$

c) $x - \frac{1}{3} > \frac{2}{3}$

$x > 1$

a) add 5 on each side   divide by 2   add $\frac{1}{3}$

3) State three values that satisfy each inequality; one integer, one fraction, and one decimal

a) $x + 3 < 5$

b) $x - 4 > 1$

c) $5x \leq 20$

a) subtract 3 on each side

$x + 3 - 3 < 5 - 3$

$x < 2$

Three of the solutions are: -1, $\frac{1}{2}$, 1.89

b) add 4 on each side

$x - 4 + 4 > 1 + 4$

$x > 5$

Three of the solutions are: 10, $7\frac{1}{2}$, 5.89

c) divide by 5 on both sides

$\frac{5}{5}x \leq \frac{20}{5}$

$x \leq 4$

Three of the solutions are: -15.34, $\frac{1}{2}$, 4

4) Melissa has \$310 in her bank account. She must maintain a minimum balance of \$600 in her bank account to avoid paying a monthly fee. How much money can Melissa deposit into her account to avoid paying bank fees?

a) Choose a variable and write an inequality to solve the problem.

b) Solve the problem

a) The variable will be x – the money to be deposited, and the inequality would be:

$310 + x \geq 600$

b) $310 + x \geq 600$

$310 - 310 + x \geq 600 - 310$

$x \geq 290$

So, Melissa has to deposit at least \$290 in order to avoid paying bank fees.

5) A water slide charges \$2 to rent an inflatable ring, and \$0.5 per ride. Iveta has \$12. How many rides can Iveta go on?

$2 + 0.5x \geq 12$

$2 - 2 + 0.5x \geq 12 - 2$

$0.5x \geq 10$

$\dfrac{0.5}{0.5}x \geq \dfrac{10}{0.5}$

$x \geq 20$ rides

4.C. Solving multi-step linear inequalities

1) Solve and check

$5x + 7 \geq 2$

Step 1 we subtract 7 on each side of the inequality.

$5x + 7 - 7 \geq 2 - 7$

$5x \geq -5$

Step 2 we divide by 5 on each side of the inequality

$\dfrac{5}{5}x \geq -\dfrac{5}{5}$

$x \geq -1$

Check

We choose any number that is greater or equal with minus 1

Let us choose 0 and substitute it instead of x in the original inequality

$5(0) + 7 \geq 2$

$0 + 7 \geq 2$

$7 \geq 2$ which is true

So, $x \geq -1$ is the solution for the inequality $5x + 7 \geq 2$

2) Solve and graph the solution

$3x + 4 \geq 6 + 2x$

Step 1 we subtract 4 on each side of the inequality

$3x + 4 - 4 \geq 6 - 4 + 2x$

$3x \geq 2 + 2x$

Step 2 we subtract $2x$ on both sides.

$3x - 2x \geq 2 + 2x - 2x$

$x \geq 2$

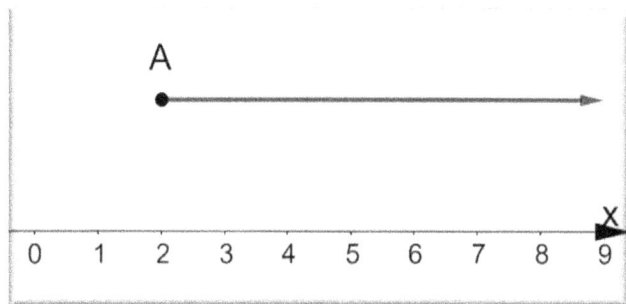

3) Solve

$\frac{2}{5}x - \frac{1}{2} > 3 + x$

Step 1 add ½ on each side of the inequality

$\frac{2}{5}x - \frac{1}{2} + \frac{1}{2} > 3 + \frac{1}{2} + x$

$\frac{2}{5}x > \frac{7}{2} + x$

Step 2 subtract x on each side of the inequality

$\frac{2}{5}x - x > \frac{7}{2} + x - x$

$x\left(\frac{2}{5} - 1\right) > \frac{7}{2}$

Step 3 we follow the order of operations rule and do the bracket $\left(\frac{2}{5} - 1\right)$

$x\left(\frac{2}{5} - \frac{5}{5}\right) > \frac{7}{2}$

$-\frac{3}{5}x > \frac{7}{2}$

Step 4 we multiply by 10 (common denominator) on each side of the inequality

$-\frac{3*10}{5}x > \frac{7*10}{2}$

$-3 * 2x > 7 * 5$

$-6x > 35$

Step 5 we multiply by -1 on each side of the inequality

The sign of the inequality changes in <

We have:

$6x < -35$

Step 6 we divide by 6 on each side of the inequality

$\frac{6}{6}x < -\frac{35}{6}$

$x < -5\frac{5}{6}$

4) Your school wants to raise money for charity. The school organizes a dance where the DJ costs $1200 and the ticket costs $8. How many tickets have to be sold to make a profit more than $1700?

a) Write an inequality to solve the problem

b) Solve and verify the solution

a) The variable x represents the number of students that buy tickets.

The inequality is:

$8x - 1200 \geq 1700$

b) Solve

Step 1 add 1200 on both sides of the inequality

$8x - 1200 + 1200 \geq 1700 + 1200$

$8x \geq 2900$

Step 2 we divide by 8 on both sides of the inequality

$\frac{8}{8}x \geq \frac{2900}{8}$

$x \geq 362$ students

Check

We choose a value of 400 for x

$8(400) - 1200 \geq 1700$

$3200 - 1200 \geq 1700$

$2000 \geq 1700$ it is true

So, $x \geq 362$ students is the solution for the inequality $8x - 1200 \geq 1700$

5) Solve

$3(x - 3) > \frac{2}{3}(3x + 6)$

Step 1 we expand the rackets by using the **distributivity property**

$3x - 9 > 2x + 4$

Step 2 we add 9 on both sides of the inequality

$3x - 9 + 9 > 2x + 4 + 9$

$3x > 2x + 13$

Step 3 we subtract $2x$ on each side of the inequality

$3x - 2x > 2x - 2x + 13$

$x > 13$

6) John is replacing the light bulbs in his house from regular to energy saver light bulbs.

A regular light bulb costs $0.6 and has an electricity cost of $0.005 per hour.

An energy saver light bulb costs $5.5 and has an electricity cost of $0.001 per hour.

For how many hours of use it is cheaper to use an energy saver light bulb than a regular light bulb?

a) Write an inequality for this problem.

b) Solve the inequality. Explain the solution in words.

a) $5.5 + 0.001x < 0.6 + 0.004x$

b) Solve it

Step 1 we subtract 5.5 on each side of the inequality

$5.5 - 5.5 + 0.001x < 0.6 - 5.5 + 0.004x$

$0.001x < -4.9 + 0.004x$

Step 2 we subtract $0.004x$ on each side of the inequality

$0.001x - 0.004x < -4.9 + 0.004x - 0.004x$

$-0.003x < -4.9$

Step 3 we multiply by -1 on each side of the inequality

$-0.003(-1)x < -4.9(-1)$

The sign of the inequality changes to >

$0.003x > 4.9$

Step 4 we divide by 0.003 on each side of the inequality

$\frac{0.003}{0.003}x > \frac{4.9}{0.003}$

$x > 1,633.3 \ hours$

The energy saver light bulb has to be used more than 1,633.3 hours to be cost effective

4.D. Linear inequalities with two variables

1) Graph the solution

$y > 3x - 1$

Step 1 we graph the line $y = 3x - 1$

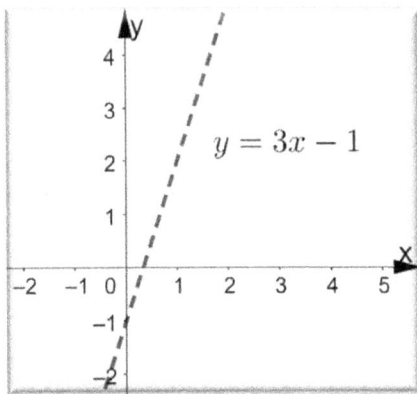

Step 2 we choose a test point, say (0,0) and see if this point satisfies the inequality.

$y > 3x - 1$

So,

$0 > 3(0) - 1$

$0 > 0 - 1$

$0 > -1$ it is true

It means that point (0,0) is part of the solution. The solution is all the points situated at the left of the line $y = 3x - 1$

We will graph the solution.

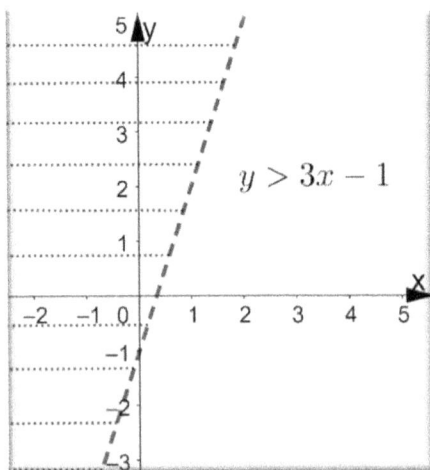

2) Which point(s) is/are in the solution region of the inequality $3x - 6y \leq 4$

a) (0,0)   b) (4,2)   c) (-2,5)   d) (3,-6)

---

a) we substitute the coordinates (0,0) in the inequality

$3(0) - 6(0) \leq 4$

$0 \leq 4$ is true, so, point (0,0) is in the solution region of the inequality $3x - 6y \leq 4$

b) we substitute the coordinates (4,2) in the inequality

$3(4) - 6(2) \leq 4$

$0 \leq 4$ is true, so, point (4,2) is in the solution region of the inequality $3x - 6y \leq 4$

c) we substitute the coordinates (-2,5) in the inequality

$3(-2) - 6(5) \leq 4$

$-6 - 30 \leq 4$

$-36 \leq 4$

is true, so, point (-2,5) is in the solution region of the inequality $3x - 6y \leq 4$

d) we substitute the coordinates (3,-6) in the inequality

$3(3) - 6(-6) \leq 4$

$9 + 36 \leq 4$

$45 \leq 4$

is not true, so, point (3,-6) is not in the solution region of the inequality $3x - 6y \leq 4$

3) The graph below, the equation of the boundary line is: $x - 3y = 6$

Determine the inequality represented by the graph.

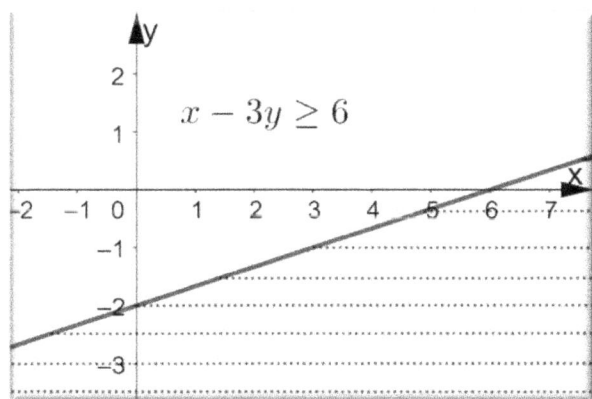

Step 1 we add $3y$ on both sides of the equation

$x - 3y = 6$

$x - 3y + 3y = 6 + 3y$

$x = 6 + 3y$

Step 2 we subtract 6 on both sides of the equation

$x - 6 = 6 - 6 + 3y$

$x - 6 = 3y$

Step 3 we divide by 3 on both sides of the equation

$\frac{1}{3}x - \frac{6}{3} = \frac{3}{3}y$

$\frac{1}{3}x - 2 = y$

$$y = \frac{1}{3}x - 2$$

Step 4 we substitute point (0,0) into the equation and use the sign $\leq$ or $\geq$ in such a way that the point is not part of the solution.

$$0 = \frac{1}{3}(0) - 2$$

$$0 = -2$$

So, we will use the sign that will not satisfy the inequality for (0,0)

$$x - 3y \geq 6$$

4) The graph below shows the solution to the inequality

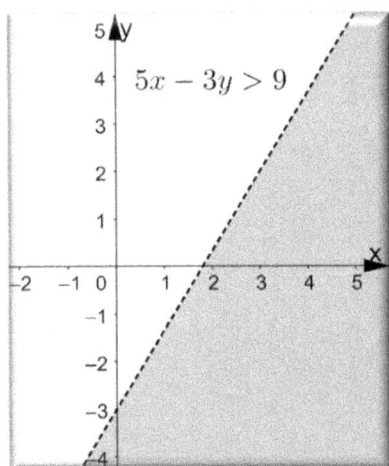

a) Explain why the boundary line is a broken line?

The left side is only bigger than the right side, not equal. The points situated on the line are not part of the solution. To show that, the line is represented boken and not continue.

b) Why the solution is beneath the line and not above?

If we choose the test point (0,0), the coordinates zero and zero substituted in the inequality, this inequality will become:

$$5(0) - 3(0) > 9$$

Or, $0 > 9$ which is not correct.

From here we have that point (0,0) is not part of the solution, so the solution is the region beneath the line, not including the point on the line.

5) The point(s) which is NOT in the solution region of the inequality $5x - 2y > 5$ are:

a) (1,3)   b) (3,2)   c) (2,-3)   d(0,1)

a) $5(1) - 2(3) > 5$

$5 - 6 > 5$

$-1 > 5$ NOT TRUE   point (1,3) is not part of the solution region

b) $5(3) - 2(2) > 5$

$15 - 4 > 5$

$11 > 5$ true

c) $5(2) - 2(-3) > 5$

$10 + 6 > 5$

$16 > 5$ true

d) $5(0) - 2(1) > 5$

$0 - 2 > 5$

$-2 > 5$ NOT TRUE point $(0,1)$ is not part of the solution region

CHAPTER 5

5.B. Adding and subtracting polynomials

1) Determine if the expressions below are a monomial, binomial or trinomial.

a) $3x - 2$,   b) 4 c) $5x^2 + 3x - 1$ d) $x^2 - 3$

a) Binomial   b) monomial   c) trinomial   d) Binomial

2) Determine the degree of the following polynomials

a) $5x^3 + 3x^2 - 2x + 4$ b) 5 c) $4x^2 - 5x + 6$

Degree of the polynomial is the highest degree of the variables.

a) The degree of $5x^3 + 3x^2 - 2x + 4$ is three

b) A constant (a number) has the degree equal with zero.

c) The degree of $5x^3y^2 + 3x^2y - 2x + 4$ is five

3) Add the polynomials:

a) $(2x^2 - 3x + 4) + (2x - 4) =$ b) $(3x^2 + 4x + 6) + (-4x - 5) =$

a) $(2x^2 - 3x + 4) + (2x - 4) = 2x^2 - 3x + 4 + 2x - 4 = 2x^2 - 3x + 2x + 4 - 4 = 2x^2 - x$

b) $(3x^2 + 4x + 6) + (-4x - 5) = 3x^2 + 4x + 6 - 4x - 5 = 3x^2 + 4x - 4x + 6 - 5 = 3x^2 + 1$

4) Subtract the polynomials

a) $(4x^2 - 3x + 2) - (3x^2 + 5x - 7) =$ b) $(2x^2 + 4x - 5) - (-3x^2 + 2x - 1) =$

a) $(4x^2 - 3x + 2) - (3x^2 + 5x - 7) = 4x^2 - 3x + 2 - 3x^2 - 5x + 7 = 4x^2 - 3x^2 - 3x - 5x + 2 + 7 = x^2 - 8x + 9$

b) $(2x^2 + 4x - 5) - 3x^2 + 2x - 1 = 2x^2 + 4x - 5 + 3x^2 - 2x + 1 =$
$= 2x^2 + 3x^2 + 4x - 2x - 5 + 1 = 5x^2 + 2x - 4$

5) Add and/or subtract the polynomials

a) $(4x^3 - 3x^2 - 2x + 1) + (2x^3 + 3x^2 - 4x - 5) =$

b) $(4x^3 - 2x + 1) - (2x^3 + 3x^2 - 4x) =$

a) $(4x^3 - 3x^2 - 2x + 1) + (2x^3 + 3x^2 - 4x - 5) = 4x^3 - 3x^2 - 2x + 1 + 2x^3 + 3x^2 - 4x - 5 =$
$= 4x^3 + 2x^3 - 3x^2 + 3x^2 - 2x - 4x + 1 - 5 = 6x^3 - 6x - 4$

b) $(4x^3 - 2x + 1) - (2x^3 + 3x^2 - 4x) = 4x^3 - 2x + 1 - 2x^3 - 3x^2 + 4x =$
$= 4x^3 - 2x^3 - 3x^2 - 2x + 4x + 1 = 2x^3 - 3x^2 + 2x + 1$

6) Add and/or subtract the polynomials
$(4x^3 + 1) - (2x^3 + 3x^2) + (5x^2 - 6x - 4) =$

$(4x^3 + 1) - (2x^3 + 3x^2) + (5x^2 - 6x - 4) = 4x^3 + 1 - 2x^3 - 3x^2 + 5x^2 - 6x - 4 =$
$= 4x^3 - 2x^3 - 3x^2 + 5x^2 - 6x + 1 - 4 = 2x^3 + 2x^2 - 6x - 3$

5.C. Multiplication of polynomials

1) $(3x^2 - 2)(x + 1) =$

$(3x^2 - 2)(x + 1) = 3x^3 + 3x^2 - 2x - 2$

2) $(x^2 + 3)(4x - 5) =$

$(x^2 + 3)(4x - 5) = 4x^3 - 5x^2 + 12x - 15$

3) $(xy - x^2 + 3)(y + 3xy - 2) =$

$(xy - x^2 + 3)(y + 3xy - 2) = xy^2 + 3x^2y^2 - 2xy - x^2y - 3x^3y + 2x^2 + 3y + 9xy - 6 = xy^2 + 3x^2y^2 + 7xy - x^2y - 3x^3y + 2x^2 + 3y - 6 =$

4) $(x - 1)(x + 1)(2xy + x + y) =$

$(x - 1)(x + 1)(2xy + x + y) = (x^2 - 1)(2xy + x + y) = 2x^3y + x^3 + x^2y - 2xy - x - y$

5) $(2x - 3)(2x + 3)(x + x^2 - x^3 + 4) =$

$(2x - 3)(2x + 3)(x + x^2 - x^3 + 4) = (2x^2 - 9)(x + x^2 - x^3 + 4) = 2x^3 + 2x^4 - 2x^5 + 8x^2 - 9x - 9x^2 + 9x^3 - 36 = -2x^5 + 2x^4 + 11x^3 - x^2 - 9x - 36$

5.D. Rationalizing the denominator, special binomial products

1) The rationalized expression of $\frac{2}{\sqrt{7}}$ is

$\frac{2}{\sqrt{7}} = \frac{2*\sqrt{7}}{\sqrt{7}*\sqrt{7}} = \frac{2*\sqrt{7}}{7}$

2) The rationalized expression of $\frac{2\sqrt{3}}{\sqrt{11}}$ is

$\frac{2\sqrt{3}}{\sqrt{11}} = \frac{2\sqrt{3}*\sqrt{11}}{\sqrt{11}*\sqrt{11}} = \frac{2\sqrt{33}}{11}$

3) The rationalized expression of $\frac{\sqrt{7}+3\sqrt{5}}{\sqrt{3}}$ is

$\frac{\sqrt{7}+3\sqrt{5}}{\sqrt{3}} = \frac{(\sqrt{7}+3\sqrt{5})*\sqrt{3}}{\sqrt{3}*\sqrt{3}} = \frac{\sqrt{7}*\sqrt{3}+3\sqrt{5}*\sqrt{3}}{\sqrt{3}*\sqrt{3}} = \frac{\sqrt{21}+3\sqrt{15}}{3}$

4) The rationalized expression of $\frac{\sqrt{5}}{\sqrt{3}+\sqrt{2}}$ is

$\frac{\sqrt{5}}{\sqrt{3}+\sqrt{2}} = \frac{\sqrt{5}*(\sqrt{3}-\sqrt{2})}{(\sqrt{3}+\sqrt{2})*(\sqrt{3}-\sqrt{2})} = \frac{\sqrt{5}*\sqrt{3}-\sqrt{5}*\sqrt{2}}{(\sqrt{3})^2-(2)^2} = \frac{\sqrt{15}-\sqrt{10}}{3-2} = \sqrt{15} - \sqrt{10}$

5) The area of a rectangle is $\sqrt{7} + 2\sqrt{5}$ and length $\sqrt{3} - 1$. The width is

$Area = (\sqrt{7} + 2\sqrt{5}) = (\sqrt{3} - 1) * width$

$width = \frac{Area}{length} = \frac{\sqrt{7}+2\sqrt{5}}{\sqrt{3}-1} = \frac{(\sqrt{7}+2\sqrt{5})*(\sqrt{3}+1)}{(\sqrt{3}-1)*(\sqrt{3}+1)} = \frac{\sqrt{7}*\sqrt{3}+\sqrt{7}+2\sqrt{5}*\sqrt{3}+2\sqrt{5}}{3-1} = \frac{\sqrt{21}+\sqrt{7}+2\sqrt{15}+2\sqrt{5}}{2}$

5.E. Applications of polynomials

1) Calculate the area of the figure below

EC=2

A $x+1$ B

$2x-3$

E C

D

$x+3$

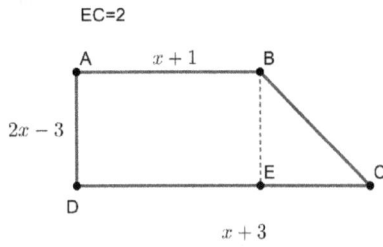

$Area_{ABED} = DE \times AD = (x+1)(2x-3) = 2x^2 - 3x + 2x - 3 = 2x^2 - x - 3$

$Area_{BEC} = \dfrac{EC \times EB}{2} = \dfrac{2(2x-3)}{2} = 2x - 3$

$Total\ Area = 2x^2 - x - 3 + 2x - 3 = 2x^2 + x - 6$

2) Calculate the area of the figure below

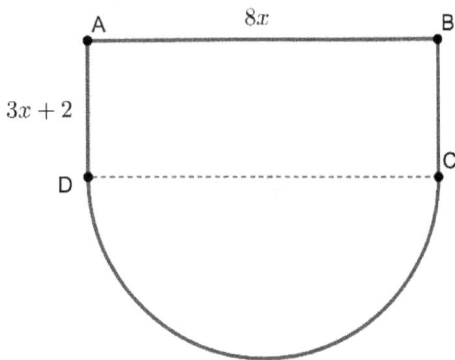

A $8x$ B

$3x+2$

D C

$Area_{ABCD} = DC \times AD = 8x(3x+2) = 24x^2 + 16x$

$Area_{semicircle} = \dfrac{\pi R^2}{2} = \dfrac{\pi(4x)^2}{2} = \dfrac{\pi(16x^2)}{2} = 8\pi x^2 \cong 25.12x^2$

$Total\ Area = 24x^2 + 16x + 25.12x^2 = 49.12x^2 + 16x$

3) Calculate the volume of the figure below

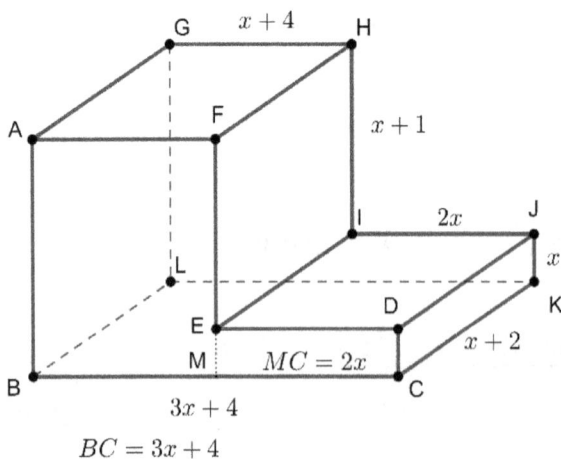

G $x+4$ H

F

A $x+1$

I $2x$ J

L x

E D K

M $MC = 2x$ $x+2$

B C

$3x+4$

$BC = 3x+4$

$Volume = A_{base} \times height = Area_{ABCDEF} \times CK$

$Area_{ABCDEF} = Area_{ABMF} + Area_{MEDC}$

$Area_{ABMF} = BM \times MF + MC \times ME$
$= (x+4)(2x+1)$
$= 2x^2 + x + 8x + 4 = 2x^2 + 9x + 4$

$Area_{MEDC} = MC \times ME = 2x(x) = 2x^2$

$Area_{ABMF} + Area_{MEDC} = 2x^2 + 9x + 4 + 2x^2 = 4x^2 + 9x + 4 = A_{base}$

$Volume = A_{base} \times height = Area_{ABCDEF} \times CK = (4x^2 + 9x + 4) \times (x + 2) = 4x^3 + 8x^2 + 9x^2 + 18x + 4x + 8 = 4x^3 + 17x^2 + 12x + 8$

4) Calculate the perimeter of the figure below

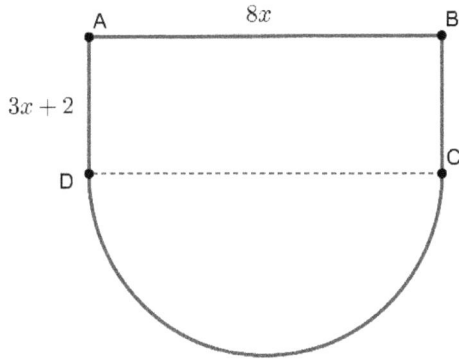

$P = 3x + 2 + 8x + 3x + 2 + \frac{2 \times \pi \times (4x)}{2} = 14x + 4 + 4x\pi =$

$14x + 12.56x + 4 = 26.56x + 4$

5) Calculate the perimeter of the figure below, BC = 9

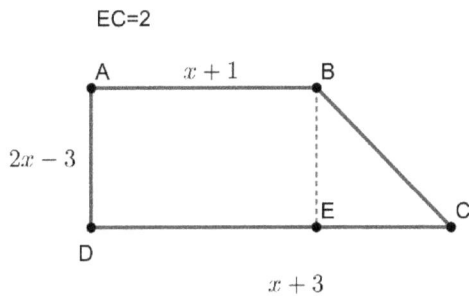

$P = x + 1 + 9 + x + 3 + 2x - 3 = 4x + 10$

5.F. Division of Polynomials

a. Long division

1) Divide: $x^2 - 3x + 4 \div x - 2$

$$
\begin{array}{r}
x - 1 \\
x - 2 \overline{\smash{\big)}\, x^2 - 3x + 4} \\
\underline{-x^2 + 2x} \\
= -x + 4 \\
\underline{x - 2} \\
= 2
\end{array}
$$

2) Divide $2x^3 - 3x^2 + 4x - 5 \div x + 4$

$$
\begin{array}{r}
2x^2 - 11x + 48 \\
\end{array}
$$

$x + 4 \enclose{longdiv}{\quad 2x^3 - 3x^2 + 4x - 5}$

$\quad\quad -(2x^3 + 8x^2)$

$\quad\quad = \ -11x^2 + \ 4x$

$\quad\quad\quad -(-11x^2 - 44x)$

$\quad\quad\quad = \quad 48x - \quad 5$

$\quad\quad\quad\quad -(48x + 192)$

$\quad\quad\quad\quad = \quad\quad -197$

3) Divide $5x^4 - 4x^3 + 3x^2 - 2x + 1 \div x^2 + 2x - 3$

$$
\begin{array}{r}
5x^2 - 14x + 46 \\
\end{array}
$$

$x^2 + 2x - 3 \enclose{longdiv}{\quad 5x^4 - 4x^3 + \ 3x^2 - 2x + 1}$

$\quad\quad -(5x^4 + 10x^3 - 15x^2)$

$\quad\quad = \ -14x^3 + 18x^2 - \ 2x$

$\quad\quad -(-14x^3 - 28x^2 - 42x)$

$\quad\quad\quad = \quad 46x^2 - 40x + \ 1$

$\quad\quad\quad\quad -(46x^2 + 92x - 138)$

$\quad\quad\quad\quad = \quad\quad -132x + 139$

4) Divide $4x^4 + 2x^2 - x + 5 \div 2x^2 - x + 4$

$$
\begin{array}{r}
2x^2 + x - 2.5 \\
2x^2 - x + 4 \;\big|\; 4x^4 - 0x^3 + 2x^2 - x + 5 \\
-(4x^4 - 2x^3 + 8x^2) \\
\hline
= \quad 2x^3 - 6x^2 - x \\
-(2x^3 - x^2 + 4x) \\
\hline
= \quad -5x^2 - 5x + 5 \\
-(-5x^2 + 2.5x - 10) \\
\hline
= \quad -7.5x + 15
\end{array}
$$

5) Divide $x^3 - 3 \div x + 2$

$$
\begin{array}{r}
x^2 - 2x + 4 \\
x + 2 \;\big|\; x^3 + 0x^2 + 0x - 3 \\
-(x^3 + 2x^2) \\
\hline
= \; -2x^2 + 0x \\
-(-2x^2 - 4x) \\
\hline
= \quad 4x - 3 \\
-(4x + 8) \\
\hline
= \; -11
\end{array}
$$

b. Synthetic Division

1) Divide $4x^3 - 3x^2 + 2x + 1 \div x - 3$

```
 3 | 4  -3   2   1
   |     12  27  87
   _____
     4   9  29  88
```

$4x^3 - 3x^2 + 2x + 1 \div x - 3 = 4x^2 + 9x + 29 + \frac{88}{x-3}$

2) Determine the quotient and remainder for $x^3 - 4x + 3 \div x + 2$

```
-2 | 1   0  -4   3
   |    -2   4   0
   _____
     1  -2   0   3
```

$x^3 - 4x + 3 \div x + 2 = x^2 - 2x + \frac{3}{x+2}$

3) Divide $2x^4 + 3x^2 - 4x + 1 \div x + 1$

```
      | 2   3   0  -4   1
-1    |    -2  -1   1   3
      _____
        2   1  -1  -3   4
```

$2x^4 + 3x^3 - 4x + 1 \div x + 1 = 2x^3 + x^2 - x - 3 + \frac{4}{x+1}$

4) Divide $x^4 - 2x^3 - x^2 + 3x - 4 \div x - 2$

```
 2 | 1  -2  -1   3  -4
   |     2   0  -2   2
   _____
     1   0  -1   1  -2
```

$x^4 - 2x^3 - x^2 + 3x - 4 \div x - 2 = x^3 - x + 1 - \frac{2}{x-2}A$

5. Divide $x^2 - 1 \div x + 5$

```
-5 | 1   0   -1
   |    -5   25
   |_____
     1  -5   24
```

$$x^2 - 1 \div x + 5 = x - 5 + \frac{24}{x+5}$$

CHAPTER 6

6.B. Factoring polynomials of the form $x^2 + bx + c$

1) Factor $x^2 + 6x + 8$

The numbers that multiplied would give us 8 are 2 and 4 respectively.
If we add 2 and 4 we get 6, the coefficient of x.
So,
$$x^2 + 6x + 8 = (x + 2)(x + 4)$$

2) Factor $x^2 - 9x + 20$

The numbers that multiplied would give us 20 are -4 and -5 respectively.
If we add -4 and -5 we get -9, the coefficient of x.
So,
$$x^2 - 9x + 20 = (x - 5)(x - 4)$$

3) Factor $x^2 - 5x - 24$

The numbers that multiplied would give us -24 are -8 and 3 respectively.
If we add -8 and 3 we get -5, the coefficient of x.
So,
$$x^2 - 5x - 24 = (x - 8)(x + 3)$$

4) Factor $x^2 + 10x + 21$

The numbers that multiplied would give us 21 are 7 and 3 respectively.
If we add 7 and 3 we get 10, the coefficient of x.
So,

$$x^2 + 10x + 21 = (x + 7)(x + 3)$$

5) Factor $x^2 - 10x + 9$

The numbers that multiplied would give us 9 are -1 and -9 respectively.
If we add -1 and -9 we get -10, the coefficient of x.
So,

$$x^2 - 10x + 9 = (x - 1)(x - 9)$$

6.C. Factoring polynomials by decomposition

1) Factor $5x^2 + 2x - 3$

We have to find:

    a) two numbers that multiplied give us "**a×c**"

    b) the same two numbers added give us the coefficient "**b**"

Here:

$$5 \times (-3) = -15$$

Remember that one integer is negative and the other positive.
So,
The sum will give us 2. One is 5, the other is -3

We will **DECOMPOSE** $5x^2 + 2x - 3$ with:

$$5x^2 + 5x - 3x - 3 = 5x(x + 1) - 3(x + 1) = (x + 1)(5x - 3)$$

2) Factor $3x^2 - 2x - 8$

We have to find:

    a) two numbers that multiplied give us "**a×c**"

    b) the same two numbers added give us the coefficient "**b**"

Here:

$$3 \times (-8) = -24 = -6 \times 4$$

Remember that one integer is negative and the other positive.
So,
The sum will give us -2. One is 4, the other is -6

We will **DECOMPOSE** $3x^2 - 2x - 8$ with:

$$3x^2 - 6x + 4x - 8 = 3x(x - 2) + 4(x - 2) = (x - 2)(3x + 4)$$

3) Factor $15x^2 - 22x + 8$

We have to find:

    a) two numbers that multiplied give us "**a×c**"

    b) the same two numbers added give us the coefficient "**b**"

Here:

$$15 \times (8) = 120 = 10 \times 12$$

Remember that one integer is negative and the other also negative.
So,
The sum will give us -22. One is -10, the other is -12

We will **DECOMPOSE** $15x^2 - 22x + 8$ with:

$$15x^2 - 10x - 12x + 8 = 5x(3x - 2) - 4(3x - 2) = (3x - 2)(5x - 4)$$

4) Factor $12x^2 + 28x - 5$

We have to find:

    a) two numbers that multiplied give us "**a×c**"

    b) the same two numbers added give us the coefficient "**b**"

Here:

$$12 \times (-5) = -60 = -2 \times 30$$

Remember that one integer is negative and the other positive.
So,

The sum will give us 28. One is -2, the other is 30

We will **DECOMPOSE** $12x^2 + 28x - 5$ with:

$12x^2 - 2x + 30x - 5 = 2x(6x - 1) + 5(6x - 1) = (6x - 1)(2x + 5)$

5) Factor $21x^2 - 8x - 45$

We have to find:

    a) two numbers that multiplied give us "**a×c**"

    b) the same two numbers added give us the coefficient "**b**"

Here:

$21 \times (-45) = -945 = -35 \times 27$

Remember that one integer is negative and the other positive.

So,

The sum will give us -8. One is -35, the other is 27

We will **DECOMPOSE** $21x^2 - 8x - 45$ with:

$21x^2 + 27x - 35x - 45 = 3x(7x + 9) - 5(7x + 9) = (7x + 9)(3x - 5)$

6.D. Factoring polynomials in difference of square form

1) Factor $x^2 - 16$

Here, 16 can be written as 4^2, so, we have: $x^2 - 16 = x^2 - 4^2 = (x - 4)(x + 4)$

2) Factor $25x^2 - 9$

Here, 9 can be written as 3^2, and 25 can be written as 5^2.

So, we have: $25x^2 - 9 = (5x)^2 - 3^2 = (5x - 3)(5x + 3)$

3) Factor $36x^2 - 49$

Here, 49 can be written as 7^2, and 36 can be written as 6^2.

So, we have: $36x^2 - 49 = (6x)^2 - 7^2 = (6x - 7)(6x + 7)$

4) Factor $2x^2 - 100$

Here, 100 can be written as 10^2, and 2 can be written as $\sqrt{2}^2$.

So, we have: $2x^2 - 100 = (\sqrt{2}x)^2 - 10^2 = (\sqrt{2}x - 10)(\sqrt{2}x + 10)$

5) Factor $9x^2 - 5$

Here, 9 can be written as 3^2, and 5 can be written as $\sqrt{5}^2$.

So, we have: $9x^2 - 5 = (3x)^2 - \sqrt{5}^2 = (3x - \sqrt{5})(3x + \sqrt{5})$

6.E. Solving a quadratic equation by Factoring

1) Solve $x^2 + 6x + 8 = 0$

We can factor the trinomial. The two integers are 4 and 2.

So,

$(x + 2)(x + 4) = 0$

$x + 2 = 0, \, or \, x + 4 = 0$

$x + 2 = 0 \Longrightarrow x = -2$

Or,

$x + 4 = 0 \Rightarrow x = -4$

2) Solve: $x^2 - 36 = 0$

This $x^2 - 36$ is a difference of squares so, it can be written as $(x - 6)(x + 6)$

So,

$(x - 6)(x + 6) = 0$

$x - 6 = 0, \, or \, x + 6 = 0$

$x - 6 = 0 \Longrightarrow x = 6$

Or,

$x + 6 = 0 \Rightarrow x = -6$

3) Solve $(x - 1)^2 = 0$

$(x - 1)^2 = (x - 1)(x - 1) = 0$

So,

$x - 1 = 0 \Rightarrow x = 1, only\ one\ solution$

4) Solve $3x^2 + 15x + 12 = 0$

We can take 3 as a common factor

$3(x^2 + 5x + 4) = 0$

$x^2 + 5x + 4 = (x + 1)(x + 4) = 0$

$x + 1 = 0, or\ x + 4 = 0$

$x + 1 = 0 \Rightarrow x = -1$

Or,

$x + 4 = 0 \Rightarrow x = -4$

5) Solve

$6x^2 + 13x - 5 = 0$

We have to find:

 a) two numbers that multiplied give us "**a×c**"
 b) the same two numbers added give us the coefficient "**b**"

Here:

$6 \times (-5) = -30 = -2 \times 15$

Remember that one integer is negative and the other positive.
So,
The sum will give us 13. One is -2, the other is 15

We will **DECOMPOSE** $6x^2 + 13x - 5$ with:

$6x^2 - 2x + 15x - 5 = 2x(3x - 1) + 5(3x - 1) = (2x + 5)(3x - 1)$

So,

$(2x + 5)(3x - 1) = 0$

$2x + 5 = 0, or\ 3x - 1 = 0$

$2x + 5 = 0 \Rightarrow x = -\frac{5}{2}$

Or,

$3x - 1 = 0 \Rightarrow x = \frac{1}{3}$

6.F. Quadratic formula

1) The number to be added to $x^2 + 5x$ to make a perfect square is:

$x^2 + 5x = x^2 + 2 \times \frac{5}{2}x + (\frac{5}{2})^2$. So, the number is $+(\frac{5}{2})^2 = \frac{25}{4}$

2) The term outside the square of expression $x^2 - 5x + a$ is:

$x^2 - 6x + a = x^2 - 2 \times \frac{6}{2}x + (\frac{6}{2})^2 - (\frac{6}{2})^2 + a = (x - 3)^2 - 9 + a$

So, the term outside the square is -9+a

3) The completed square form of $x^2 + 6x - 4$ is

$x^2 + 6x - 4 = x^2 + 2 \times 3x + 9 - 9 - 4 = (x + 3)^2 - 13$

4) The completed square form of $ax^2 + 2x - 4, a \neq 0$ is $a(x + \frac{1}{a})^2 - (\frac{1+4a}{a})$

$ax^2 + 2x - 4 = a\left(x^2 + 2 \times \frac{1}{a}x + \left(\frac{1}{a}\right)^2 - \left(\frac{1}{a}\right)^2\right) - 4 = a(x + \frac{1}{a})^2 - \frac{1}{a} - 4 = a(x + \frac{1}{a})^2 - (\frac{1+4a}{a})$

5) To form a perfect square k in $x^2 + kx + 3, k = -6.25$

$x^2 + 5x + k = x^2 + 2 \times \frac{5}{2}x + (\frac{5}{2})^2 - (\frac{5}{2})^2 + k = (x + \frac{5}{2})^2 - \frac{25}{4} + k$

So,

$\frac{25}{4} + k = 0$

$k = -6.25$

The Discriminant $b^2 - 4ac$

1) Find the number of real solutions the equation $x^2 - 4x + 7 = 0$ has.

$b^2 - 4ac = (-4)^2 - 4(1)(7) = 16 - 28 = -12 < 0$

the equation will have **NO** REAL solution

2) Find the number of real solutions the equation $2x^2 - 3x + 4 = 0$ has.

$b^2 - 4ac = (-3)^2 - 4(2)(4) = 9 - 32 = -23$

the equation will have **NO** REAL solution

3) Find the number of real solutions the equation $3x^2 + 4x - 1 = 0$ has.

$b^2 - 4ac = 4^2 - 4(3)(-1) = 16 + 12 = 28 > 0$

the equation will have **two** solutions

4) Find the number of real solutions the equation $x^2 - 4x + 4 = 0$ has.

$b^2 - 4ac = (-4)^2 - 4(1)(4) = 16 - 16 = 0$

the equation will have **one** solution

5) Find the number of real solutions the equation $5x^2 + 4x - 9 = 0$ has.

$b^2 - 4ac = (4)^2 - 4(5)(-9) = 16 + 180 = 196 > 0$

the equation will have **two** solutions

CHAPTER 7

7. A. a. Determine: if a relation is a function, the values of a function, the range

1) For each value of the domain there is only one value that belongs to the range. For domain value of 3, there are two values that belong to the range: 3, and 12. (3,3) and (3,-12). Not a function

{(-4, 8), (4, 1), (-2, 3), (0, -12), (1,2),(2, 3)} is a function

2) For each value of the domain there is only one value that belongs to the range.

3) For x=2 ; $f(2) = 5(2) + 3 = 10 + 3 = 13$

For x=1 ; $f(1) = 3(1^2) - 3(1) + 2 = 3(1) - 3 + 2 = 3 - 3 + 2 = 2$

For s=2 ; $G(2) = \frac{2^2 - 3(2) + 7}{2 + 3} = \frac{4 - 6 + 7}{5} = \frac{5}{5} = 1$

4) g(t)

For t=-1; $G(-1) = 3 - 2(-1) = 3 + 2 = 5$

For t=-2 ; $G(-2) = 3 - 2(-2) = 3 + 4 = 7$

For t=3 ; $G(3) = 3 - 2(3) = 3 - 6 = -3$

F(x)

For x=-2 ; $F(-2) = (-2)^2 - 5(-2) + 1 = 4 + 10 + 1 = 15$

For x=0 ; $F(0) = (0)^2 - 5(0) + 1 = 0 - 0 + 1 = 1$

For x=3 ; $F(3) = (3)^2 - 5(3) + 1 = 9 - 15 + 1 = 7 \, not - 4$

H(c)

For c=-3 ; $H(-3) = \frac{(-3)^2 - 2(-3)}{-3 + 2} = \frac{9 + 6}{-1} = -15$

For c=0 ; $H(0) = \frac{(0)^2 - 2(0)}{0 + 2} = \frac{0}{2} = 0$

For c=3 ; $H(3) = \frac{(3)^2 - 2(3)}{3 + 2} = \frac{9 - 6}{5} = \frac{3}{5}$

7.A. b. Linear and quadratic functions and their graphs

1) Incorrect

The difference between each consecutive value belonging to the domain should be always the same. The difference between x=9 and x=7 is two not one as it should be. The difference between each consecutive value belonging to the range should be always the same. The difference between y=7 and y=5 is two, the difference between y=8 and y=7 is one, the difference between y=10 and y=8 is two again.

2) Incorrect

The difference between each consecutive value belonging to the domain should be always the same. The difference between x=2 and x=2 is zero not one as it should be.

3) The graph does not represents a linear function.

4) $f(x) = 2x^2 - 3x + 4$ represents a quadratic function.
$f(x) = x^2 - 3x$ represents a quadratic function
$f(x) = x^4 - 3x + 4$ doesn't represent a quadratic function

7.A. c.   Inverse functions and their graphs

1) $y = 3x - 5$ so, we interchange x with y and have:
$x = 3y - 5$; + 5 each side
$x + 5 = 3y$; divide with 3 both sides
$\dfrac{x+5}{3} = y$; so $f^{-1}(x) = \dfrac{x+5}{3}$

2) $y = \dfrac{3}{2x+4}$; so, we interchange x with y and have

$x = \dfrac{3}{2y+4}$; cross multiply and have

$x(2y + 4) = 3$; use the distributivity property and have
$2xy + 4x = 3$; subtract 4x from both sides
$2xy = 3 - 4x$; divide both sides with 2x
$y = \dfrac{3-4x}{2x}$ so $f^{-1}(x) = \dfrac{3-4x}{2x}$

3)

$y = \dfrac{\sqrt{x-1}}{3}$; so, we interchange x with y and have

$x = \dfrac{\sqrt{y-1}}{3}$; multiply with 3 both sides

$3x = \sqrt{y-1}$ square both sides
$9x^2 = y - 1$; add 1 in both sides
$9x^2 + 1 = y$ so $f^{-1}(x) = 9x^2 + 1$

4) $y = \dfrac{3x}{x+3}$; so, we interchange x with y and have

$x = \frac{3y}{y+3}$; cross multiply and have

$x(y + 3) = 3y$; use the distributivity property and have

$xy + 3x = 3y$; subtract xy in each side

$3x = 3y - xy$ Take y as common factor

$3x = y(3 - x)$; divide with (3-x) in both sides

$\frac{3x}{3 - x} = y$ so $f^{-1}(x) = \frac{3x}{3 - x}$

5) $y = \frac{1}{5x+3}$; so, we interchange x with y and have

$x = \frac{1}{5y+3}$; cross multiply and have

$x(5y + 3) = 1$; use the distributivity property and have

$5xy + 3x = 1$; subtract 3x in both sides

$5xy = 1 - 3x$; divide by 5x both sides

$y = \frac{1-3x}{5x}$; so $f^{-1}(x) = \frac{1-3x}{5x}$,

7.B. Polynomial Functions

1) What is the degree of the function: $f(x) = x(x + 3) - x^2 + 4$?

$f(x) = x(x + 3) - x^2 + 4 = x^2 + 3x - x^2 + 4 = 3x + 4$

The degree is 1

2) Analyze the concavity and the inflection point of $f(x) = x^3$

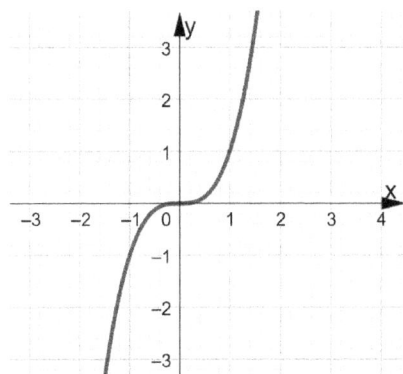

For values of x less than zero, the concavity is down.

For values of x greater than zero, the concavity is up.

The inflection point is for x equal zero.

3) Find the zeros for the function $f(x) = 2x^2 + 3x - 4$

We will use the quadratic formula.

$$x = \frac{-b \pm \sqrt{b^2 - 4ac}}{2a} = \frac{-3 \pm \sqrt{9 - 4(2)(-4)}}{2 \times 2} = \frac{-3 \pm \sqrt{9 + 32}}{4} = \frac{-3}{4} \pm \frac{\sqrt{41}}{4}$$

4) Find the zeros for the function $f(x) = x^2 + 2x - 3$

We will use the quadratic formula.

$$x = \frac{-b \pm \sqrt{b^2 - 4ac}}{2a} = \frac{-2 \pm \sqrt{4 - 4(1)(-3)}}{2 \times 1} = \frac{-2 \pm \sqrt{4 + 12}}{2} = \frac{-2}{2} \pm \frac{4}{2} = -1 \pm 2$$

$x = 1 \ and \ x = -3$

5) Find the zeros for the function $f(x) = x^2 + 2x + 7$

We will use the quadratic formula.

$$x = \frac{-b \pm \sqrt{b^2 - 4ac}}{2a} = \frac{-2 \pm \sqrt{4 - 4(1)(7)}}{2 \times 1} = \frac{-2 \pm \sqrt{4 - 28}}{2} = \frac{-2}{2} \pm \frac{\sqrt{-24}}{2} =$$

The discriminant $b^2 - 4ac$ is less than zero so, there will be NO zeros or the graph will not intersect x axis.

7.C. Piecewise functions

Determine if the expressions below represent piecewise functions.

1) $f(x) = 2x^2 + 5x - 3 , x \in R$

NO

2) $f(x) = \begin{cases} 2x \ for \ x < 0 \\ 3 \ for \ x \geq 0 \end{cases}$

YES

3) $f(x) = 4x + 3, x \in R$

NO

4) $f(x) = 3\sin(x - 2) + 3, x \in R$

NO

5) $f(x) = \sqrt{x - 1}$

NO

7.D. Trigonometric functions

1) Correct

One radian is the measure of an angle subtended at the center of a circle by an arc which is equal in length to the radius of the circle.

2) Incorrect

$\pi \, radians = 180^0$

3) Correct

In Quadrant 1 when α is between zero and 90 degrees or $\frac{\pi}{2}$, sin (α) is positive. See below.

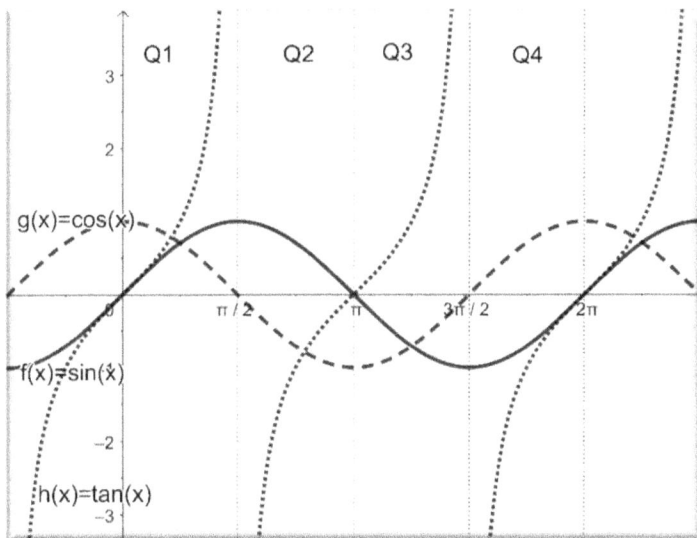

4) Correct

If $\cos(\alpha_1) = k, k \geq 0$ the other value of α that is solution of the equation is:$\alpha_2 = 2\pi - \alpha_1$ (radians). See below.

$f(x) = \cos(x)$

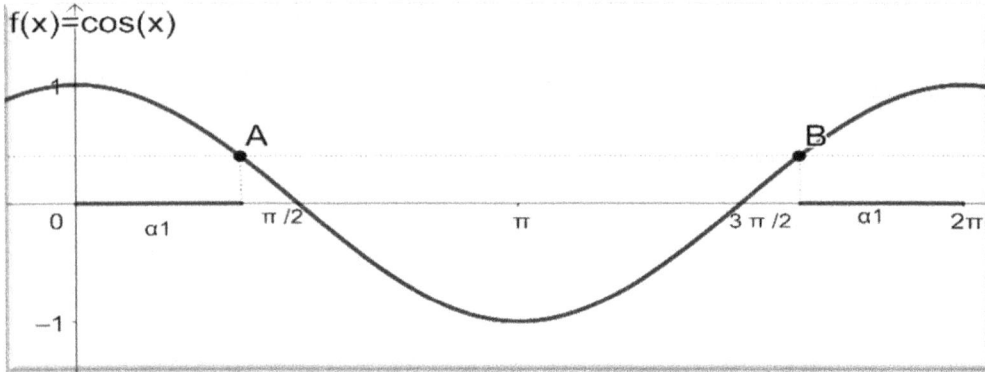

5) Incorrect

In a right-angle triangle $\cos(\alpha) = \dfrac{adjacent}{hypotenuse}$

The values for the special angles in a right-angle triangle are given below

	30^0	60^0	45^0
Sin(Φ)	$\dfrac{1}{2}$	$\dfrac{\sqrt{3}}{2}$	$\dfrac{\sqrt{2}}{2}$
Cos(Φ)	$\dfrac{\sqrt{3}}{2}$	$\dfrac{1}{2}$	$\dfrac{\sqrt{2}}{2}$
Tan(Φ)	$\dfrac{\sqrt{3}}{3}$	$\sqrt{3}$	1

7.E. a. Graphing sine and cosine functions

1) Correct

For each of values of α, the values of sin (α) are:

α	0	$\frac{\pi}{6} = 30^0$	$\frac{\pi}{2} = 90^0$	$\frac{\pi}{3} = 60^0$	π
sin (α)	0	$\frac{1}{2}$	1	$\frac{\sqrt{3}}{2}$	0

2) Incorrect

The graph of $\sin(\alpha)$ is:

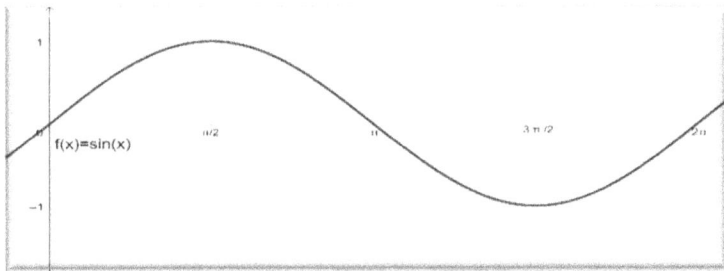

3) Correct

The minimum of $\cos(\alpha)$ is -1

4) Incorrect

The maximum of $\sin(\alpha)$ is +1   A

5) Incorrect

The graph of $\cos(\alpha)$ is:

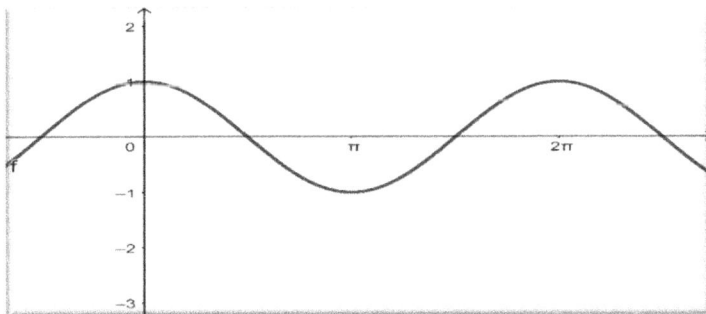

6) Incorrect

For each of values of α, the values of cos (α) are:

α	0	$\frac{\pi}{6} = 30^0$	$\frac{\pi}{2}$	$\frac{\pi}{3} = 60^0$	π

cos (α)	1	$\dfrac{\sqrt{3}}{2}$	0	$\dfrac{1}{2}$	-1

7.E. b. Graphing tangent and cotangent functions

1) Incorrect

The tangent function is defined for angle 180^0 (= 0) and *undefined for* 270^0 S

2) Correct

The tangent graph looks like the one below for $-\frac{\pi}{2}<\alpha<\frac{\pi}{2}$

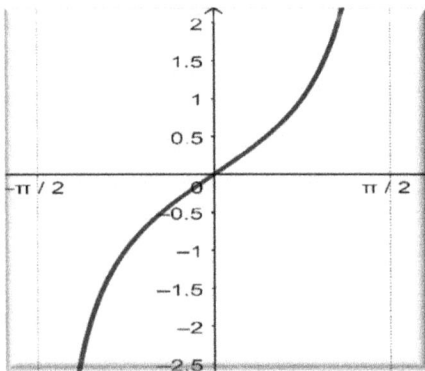

3) Correct

The cotangent function is zero for
$\frac{\pi}{2}+k\pi, k\ integer$

4) Incorrect

The tangent function is undefined for
$\frac{\pi}{2}+k\pi, k\ integer$ A

5) Correct

For 60^0 the tangent is $\sqrt{3}$

6) Correct

Tangent of 45^0 is 1

7) Incorrect

Cotangent of 30^0 is $\sqrt{3}$ T

8) Incorrect

The tangent function does not have an amplitude   U

9) Correct

Cotangent of 45^0 is 1

10) Incorrect

The formula of tangent in terms of sine and cosine is $\tan(\alpha) = \frac{\sin(\alpha)}{\cos(\alpha)}$ R

7.F.   Inverse Trigonometric Functions

1) Correct

Inverse trig functions do the opposite of the "regular" trig functions.

2) Incorrect

The angle α in right angle triangle PSQ is: 53.2^0

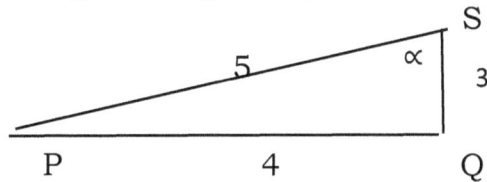

$\sin \alpha = \frac{4}{5} = 0.8$; *so* $\alpha = \sin^{-1}(0.8) = 53.2^0$

3) Incorrect

The angle α in right angle triangle PSQ is: 36.8^0

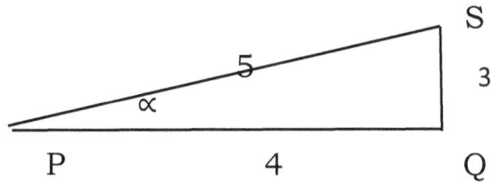

$\cos \alpha = \frac{4}{5} = 0.8$; *so* $\alpha = \cos^{-1}(0.8) = 36.8^0$

4) Correct

$Tan^{-1}\left(\sqrt{3}\right)$ is 60^0

5) Incorrect

The domain of $f(x) = \sin^{-1}(x)$ is $[-1,1]$

7.G.   Graphs of Inverse Trigonometric Functions

1) Correct

The domain of $f(x) = cos^{-1}(x)$ is $[-1,1]$

2) Incorrect

The value of $f(1) = cos^{-1}(1)$ is 0

3) Incorrect

The value of $f(0) = cos^{-1}(0)$ is $\frac{\pi}{2}$

7.H. Introduction to logarithms

Determine the answer.

1) The expression $\frac{1}{3}\log_6 a + 5\log_6 b - 7\log_6 c$ will become $\log_6(\frac{a^{\frac{1}{3}}*b^5}{c^7})$; $a, b, c > 0$

2) The expression $\log_3 5^7$ will become $7 \times \log_3 5$

3) The expression $\log_3 63 + \log_3 5 - \log_3 35 = \log_3 \frac{63*5}{35} = \log_3 \frac{63}{7} = \log_3 9 = 3$

4) The single logarithm of expression

$3(\log_5 a + \log_5 b) - 2\log_5 b = 3\log_5 a + 3\log_5 b$

$$- 2\log_5 b = \log_5 a^3 + \log_5 b^3 - \log_5 b^2 = \log_5 \frac{a^3 b^3}{b^2} = \log_5 a^3 b \,;\, a, b > 0$$

5) The expression $\log_3(\frac{a^2}{b^4})$ in terms of $\log_3 a$ and $\log_3 b$ is $2\log_3 a - 4\log_3 b$; $a, b > 0$

7.H.b. Exponential and logarithmic equations

Determine which answer is correct.

1) The solution of the equation $620 = 2 * 3^{x+1}$ is $x = \log_3 310 - 1$

$620 = 2 * 3^{x+1}$

$310 = 3^{x+1}$

$\log_3 310 = \log_3 3^{x+1}$

$\log_3 310 = (x + 1) * \log_3 3$

$\log_3 310 - 1 = x$

2) The solution of the equation $3 = \log_5 x + \log_5(x - 3)$ is:

$3 = \log_5 x + \log_5(x - 3) , x > 3$

$3 * \log_5 5 = \log_5 x(x - 3)$

$\log_5 5^3 = \log_5 x(x - 3)$

$5^3 = x(x - 3)$

$125 = x^2 - 3x$

$x^2 - 3x - 125 = 0$

$x = \frac{-(-3)\pm\sqrt{9-4(1)(-125)}}{2} = \frac{3\pm\sqrt{9+500}}{2} = \frac{3}{2} \pm \frac{\sqrt{509}}{2} = 1.5 \pm \frac{22.56}{2}$

$x_1 = 1.5 + 11.28 = 12.78$

$x_2 = 1.5 - 11.28 = -9.78$ NOT A SOLUTION

3) The solution of the equation $\log_5(3x + 9) - \log_5(x + 3) = \log_5(x - 4)$ is $x = 37$; $x > 4$

$\log_5(3x + 9) - \log_5(x + 3) = \log_5(x - 4)$

$\log_5 \frac{3x+9}{x+3} = \log_5(x - 4)$

$\frac{3(x+3)}{x+3} = x - 4$

$3 = x - 4$

$x = 7$

4) The solution of the equation $3^{x+3} = 7^{2x+5}$ is $x = -0.953$

$3^{x+3} = 7^{2x+5}$

$\ln 3^{x+3} = \ln 7^{2x+5}$

$(x + 3) \ln 3 = (2x + 5) \ln 7$

$x \ln 3 + 3 \ln 3 = 2x \ln 7 + 5 \ln 7$

$3 \ln 3 - 5 \ln 7 = 2x \ln 7 - x \ln 3$

$x(2 \ln 7 - \ln 3) = 3 \ln 3 - 5 \ln 7$

$x = \frac{3(1.0986)-5(1.9549)}{2(1.9549)-1.0986} = \frac{3.2958-5.977}{3.9098-1.0986} = \frac{-2.68}{2.811} = -0.953$

5) The solution of the equation $\log(x^2 + x - 6) - \log(x + 3) = 1$ is $x = 192$

$\log(x^2 + x - 6) - \log(x + 3) = 1$

$\log \frac{x^2+x-6}{x+3} = \log 10$

$\frac{x^2+x-6}{x+3} = 10$

$\frac{(x-2)(x+3)}{x+3} = 10$

$x - 2 = 10$

$x = 12$

7.I. Exponential and logarithmic functions

1) Represent the exponential function $f(x) = 4^x$

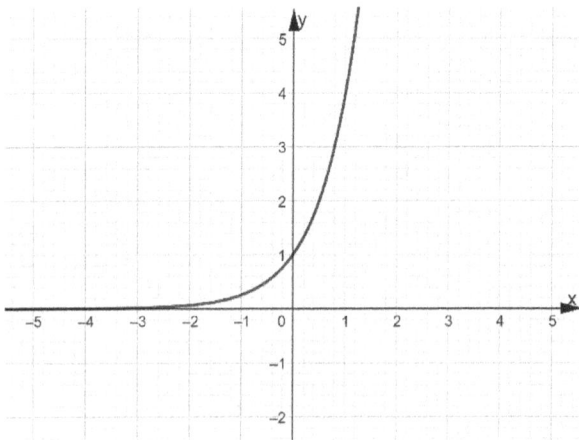

2) Represent the logarithmic function $f(x) = \log_2 x$.

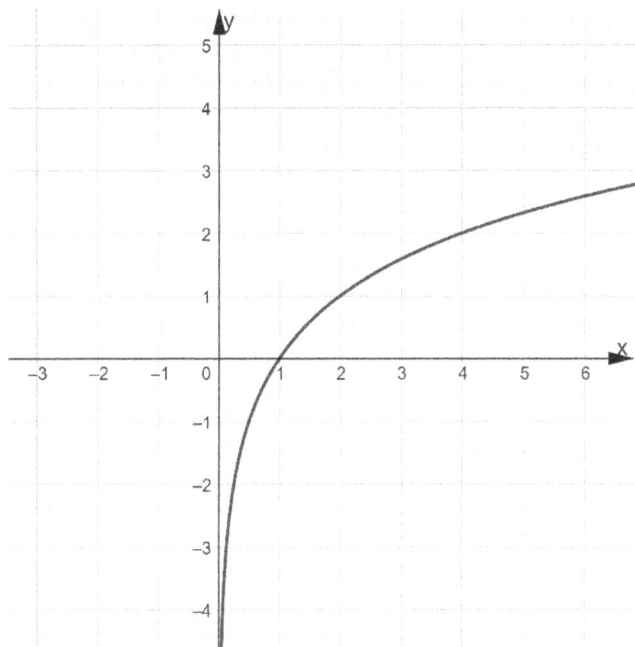

3) Represent the exponential functions $f(x) = 2^x$ and $f(x) = (\frac{1}{2})^x$ on the same graph. What do you notice?

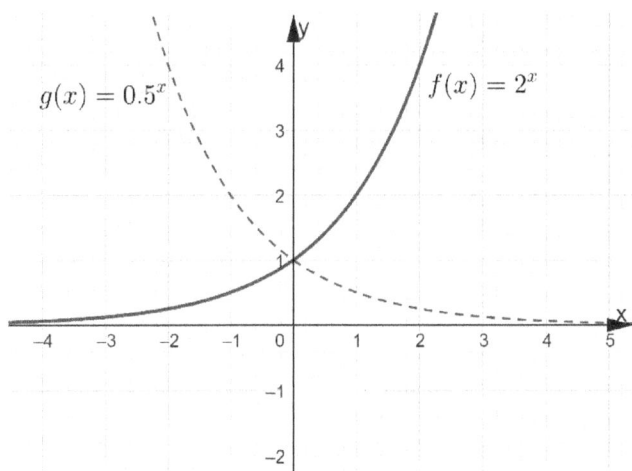

$g(x) = 0.5^x$

$f(x) = 2^x$

The graphs are symmetric to the y axis

4) Represent the logarithmic functions $f(x) = \log_2 x$ and $\log_{\frac{1}{2}} x$. What do you notice?

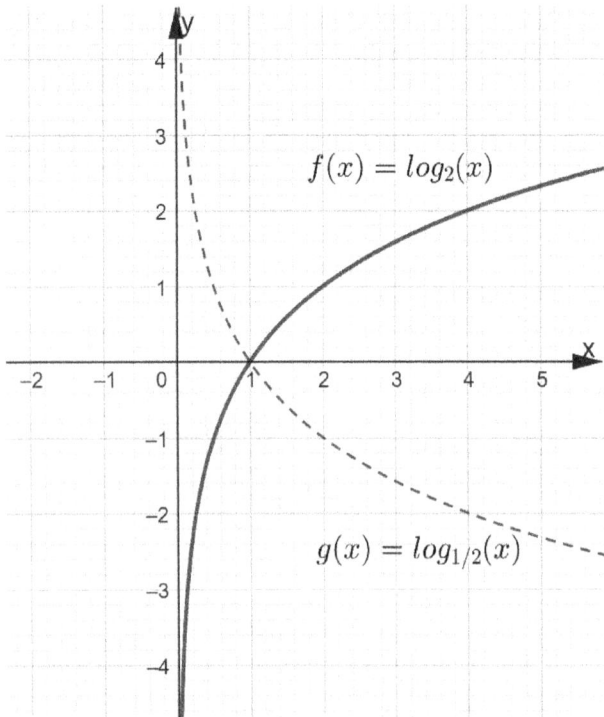

The graphs are symmetric to the x axis

$f(x) = log_2(x)$

$g(x) = log_{1/2}(x)$

5) Represent the exponential function $f(x) = 5^x$ *and* the logarithmic function $log_5 x$ on the same graph. What do you notice?

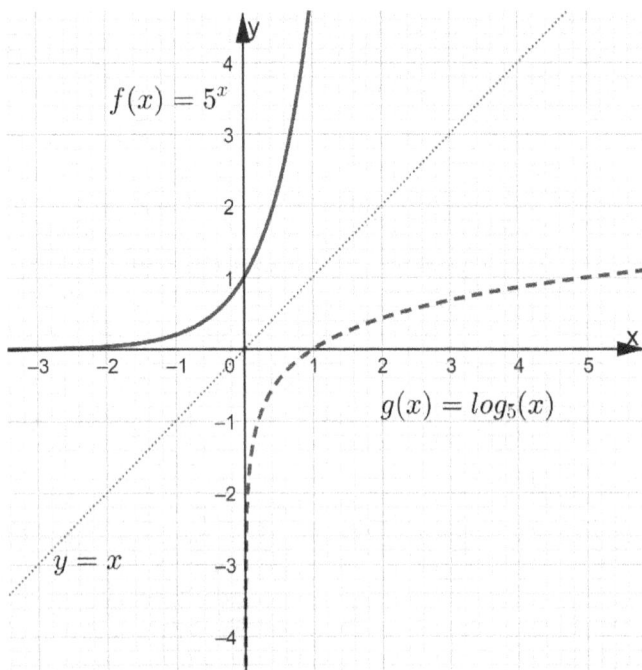

The graphs are symmetric to the y=x line

$f(x) = 5^x$

$g(x) = log_5(x)$

$y = x$

8.B. Vertical and Horizontal

Translations

Translate the graph below.

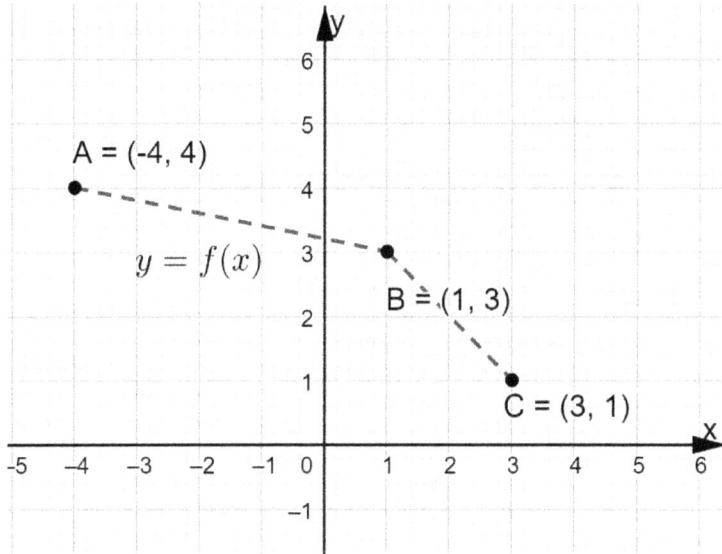

A = (-4, 4)

$y = f(x)$

B = (1, 3)

C = (3, 1)

1) 3 units up

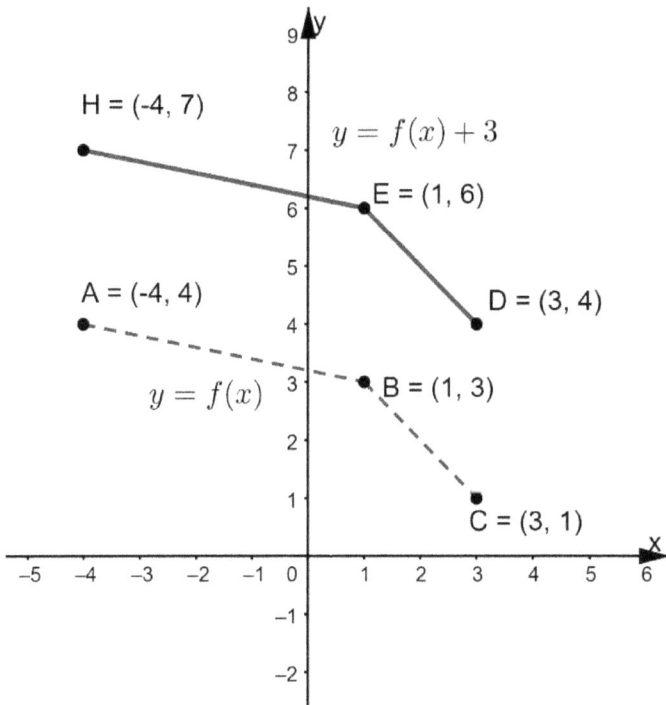

2) 2 units to the left

3) 1 unit down and 3 units to the right

4) 3 units up 2 units to the left

5) 2 units down and 4 units to the right

1) 3 units up

H = (-4, 7)

$y = f(x) + 3$

E = (1, 6)

A = (-4, 4)

D = (3, 4)

$y = f(x)$

B = (1, 3)

C = (3, 1)

2) 2 units to the left

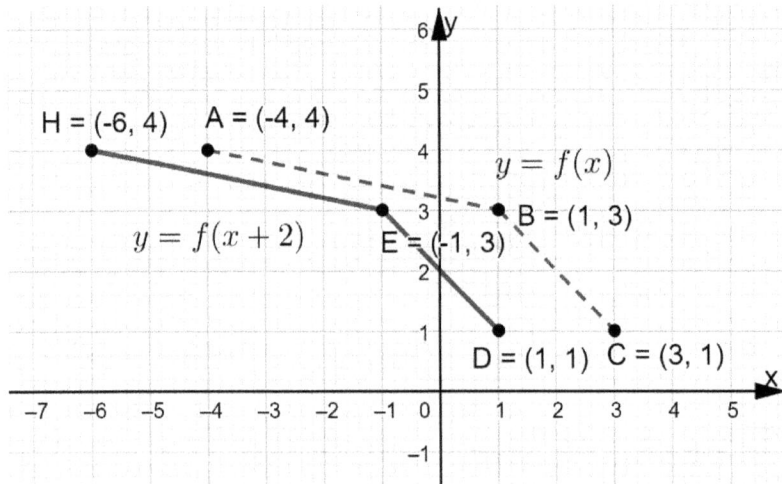

H = (-6, 4)   A = (-4, 4)

$y = f(x)$

$y = f(x + 2)$

E = (-1, 3)   B = (1, 3)

D = (1, 1)   C = (3, 1)

3) 1 unit down and 3 units to the right

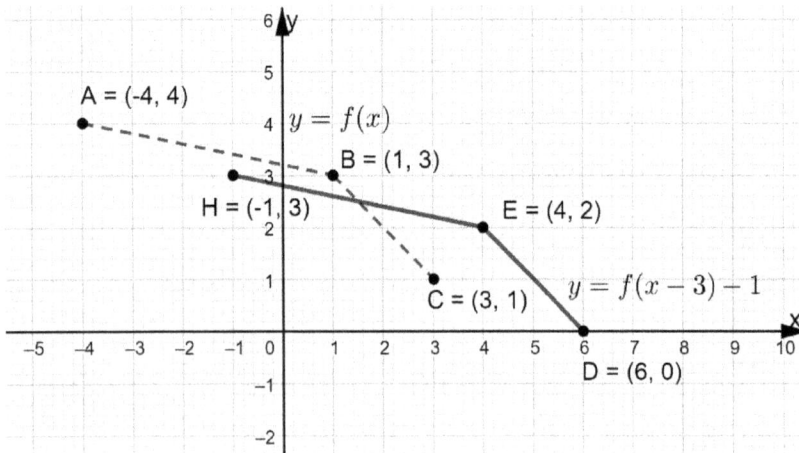

A = (-4, 4)

$y = f(x)$

B = (1, 3)

H = (-1, 3)   E = (4, 2)

C = (3, 1) $y = f(x - 3) - 1$

D = (6, 0)

4) 3 units up 2 units to the left

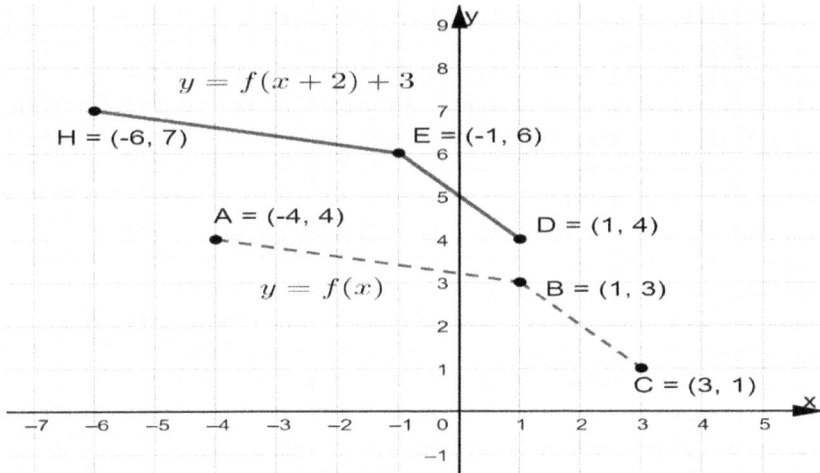

$$y = f(x + 2) + 3$$

H = (-6, 7)

E = (-1, 6)

A = (-4, 4)

D = (1, 4)

$$y = f(x)$$

B = (1, 3)

C = (3, 1)

5) 2 units down and 4 units to the right

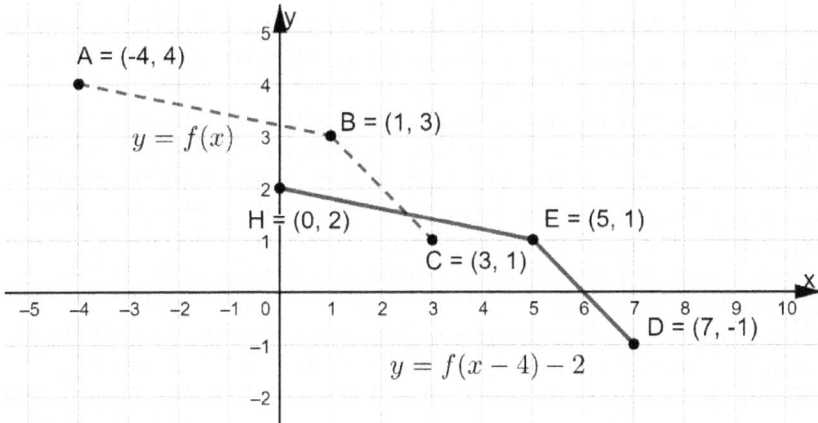

A = (-4, 4)

$$y = f(x)$$

B = (1, 3)

H = (0, 2)

E = (5, 1)

C = (3, 1)

D = (7, -1)

$$y = f(x - 4) - 2$$

8.C. Reflections

Reflect the graph below.

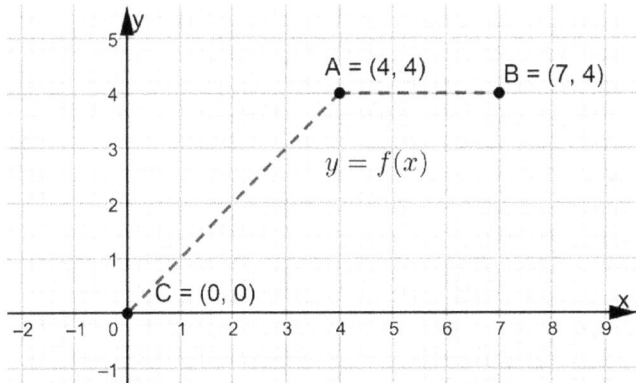

A = (4, 4)   B = (7, 4)

$$y = f(x)$$

C = (0, 0)

1) On x axis

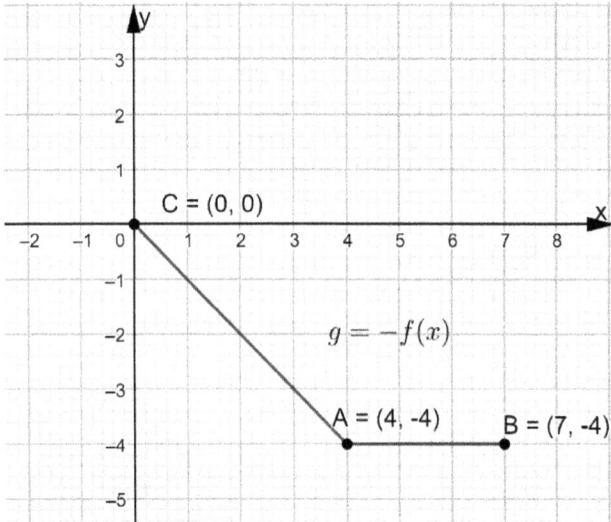

C = (0, 0)

$g = -f(x)$

A = (4, -4)   B = (7, -4)

2) On y axis

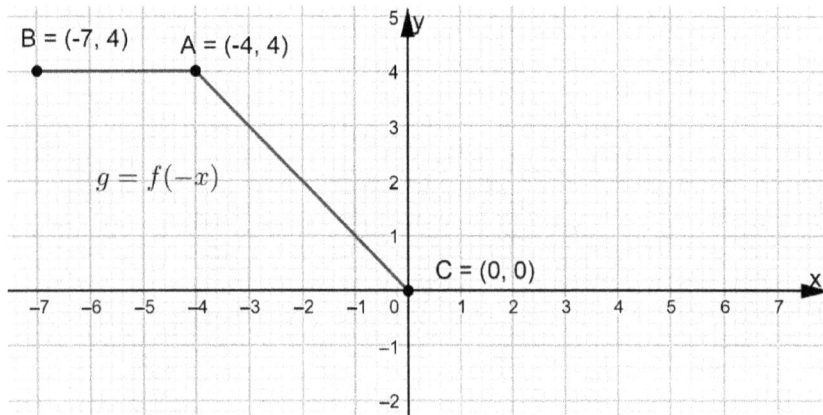

B = (-7, 4)   A = (-4, 4)

$g = f(-x)$

C = (0, 0)

3) On x and y axis

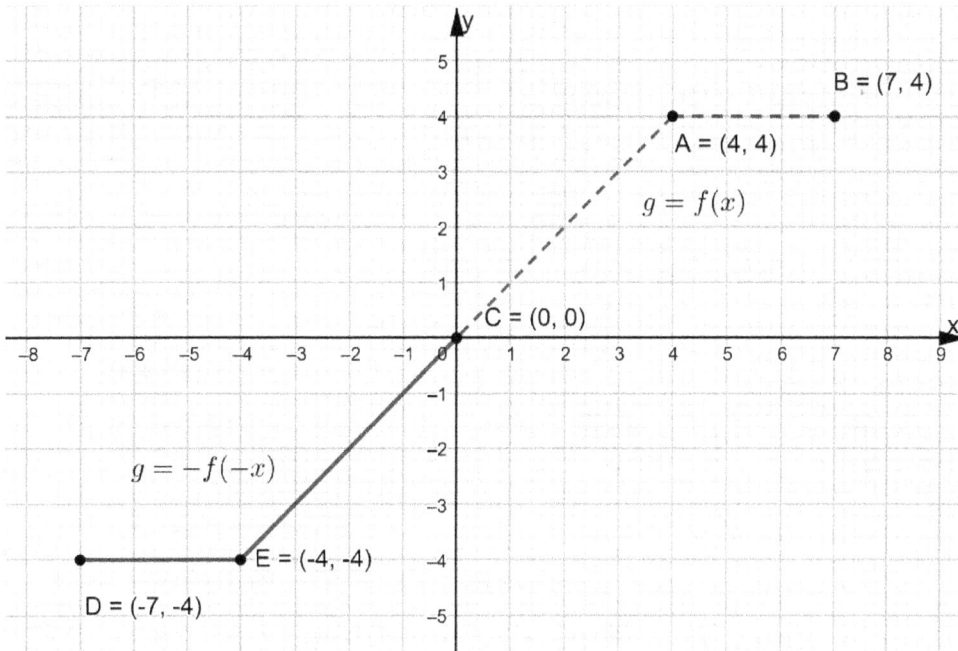

8.D. Vertical expansion or compression

1) Sketch $h(x) = -\frac{1}{2}(x + 2)^2 + 3$

$y = a(x + h)^2 + k$

a=-1/2; b=-2; c=3

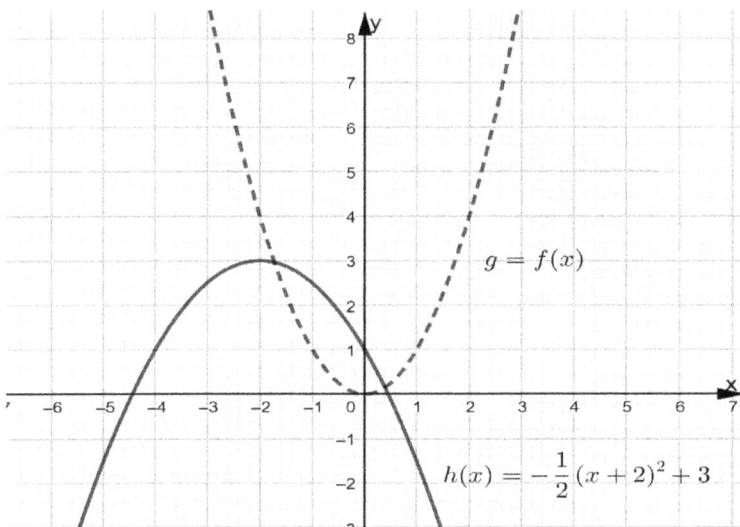

2) Sketch $y = h(x) = -2(x - 4)^2 + 1$

Here, a=-2; b=4; c=1

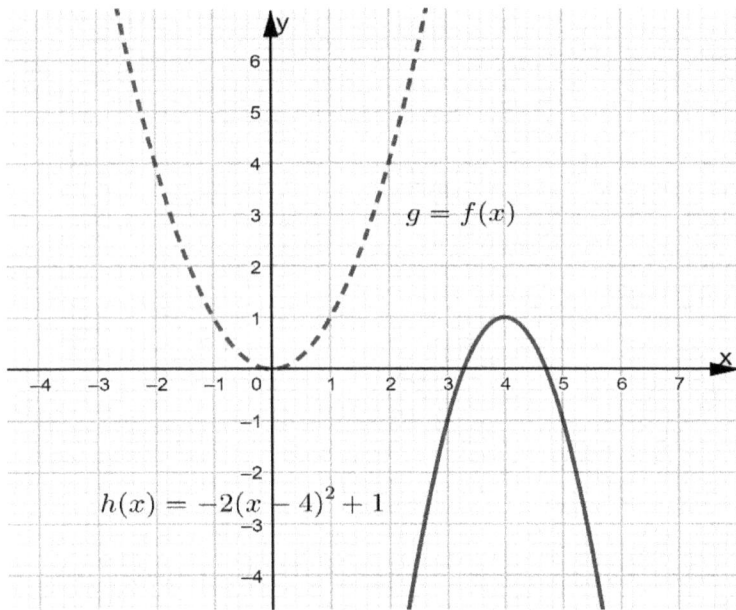

$$h(x) = -2(x - 4)^2 + 1$$

3) Sketch $y = h(x) = 3(x + 2)^2 - 4$

Here, a=3; b=-4; c=-4

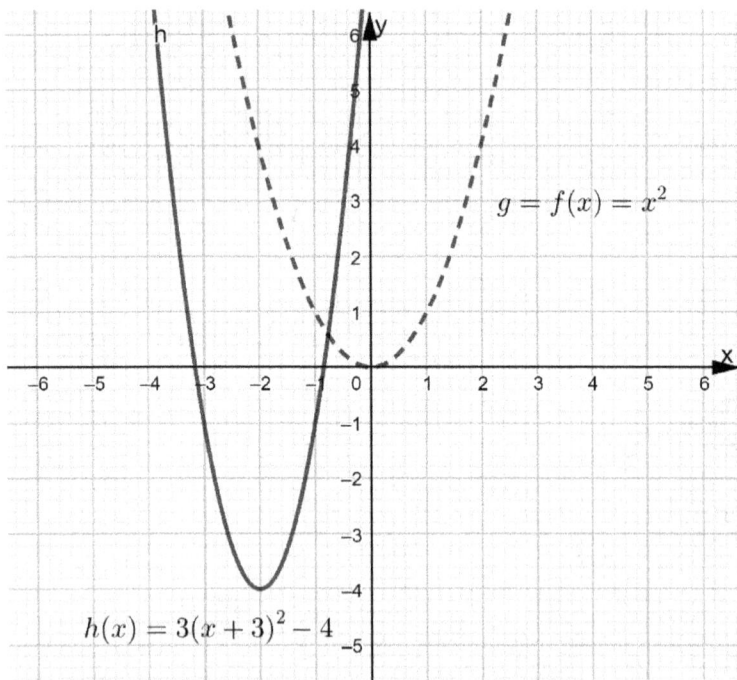

$$h(x) = 3(x + 3)^2 - 4$$

4) Sketch $y = h(x) = 3(x + 2)^2 - 4$

Here, a=3; b=-2; c=-4

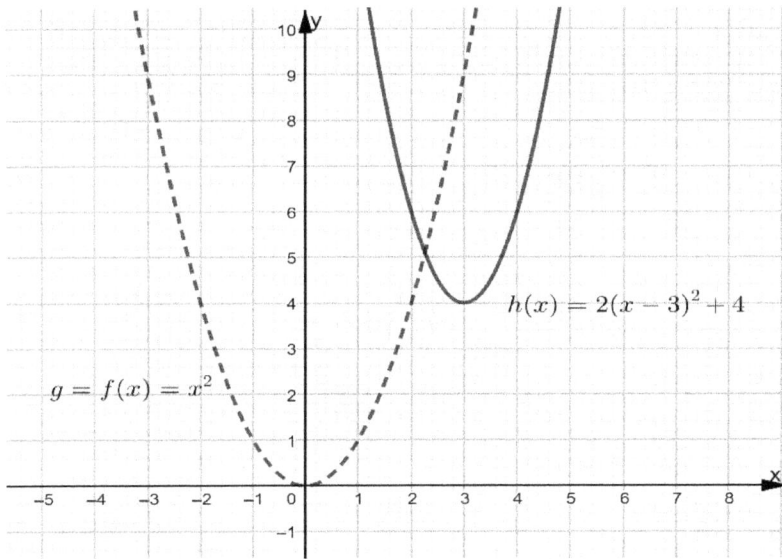

5) Sketch $y = h(x) = \frac{1}{3}(x+3)^2 + 4$

Here, a=1/3 ; b=-3; c=4

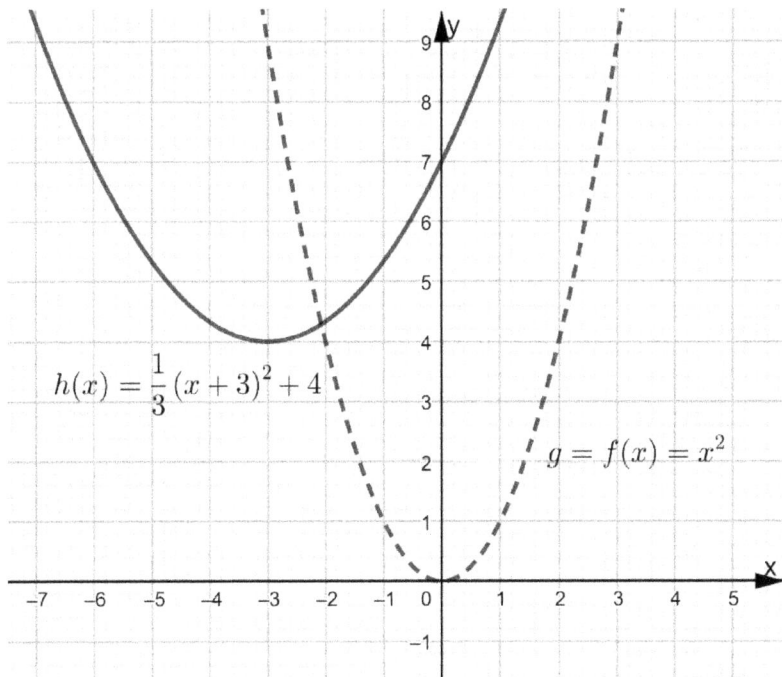

CHAPTER 9

9.A. System of linear equations using two equations and two unknowns.

b. Solving the system of equations by substitution

1) Find the solution of the system of equations and check

$$\begin{cases} y = -2x + 1 \\ x - y = 2 \end{cases}$$

$y = -2x + 1$

$x - (-2x + 1) = 2$

$x + 2x - 1 = 2$

$3x - 1 = 2$

$3x = 3$

$x = 1$

$x - y = 2$

$1 - y = 2$

$x - 2 = y$

$-1 = y$

CHECK

$y = -2x + 1$

$-1 = -2(1) + 1$

$-1 = -1$

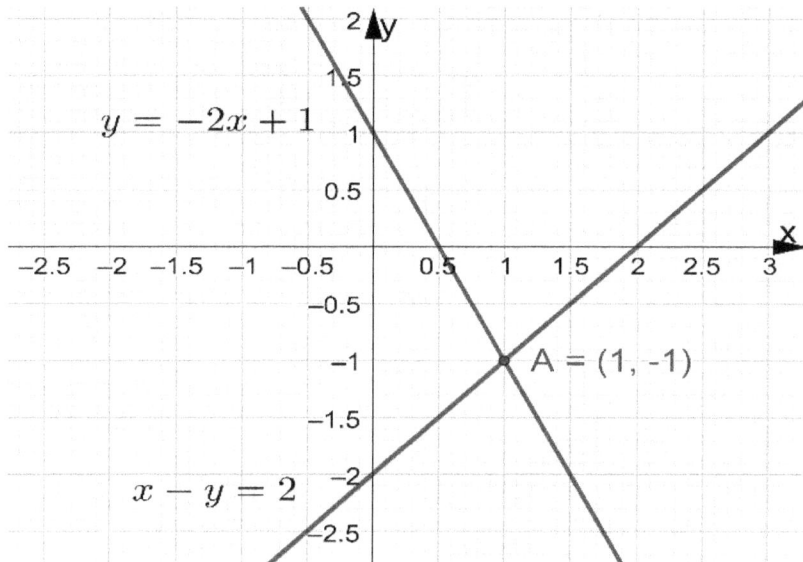

$y = -2x + 1$

$x - y = 2$

$A = (1, -1)$

2) Find the solution of the system of equations

$$\begin{cases} 3x + 4y = -5 \\ 2x - 2y = -4 \end{cases}$$

We divide the second equation by 2. We will have:

$$\begin{cases} 3x + 4y = -5 \\ x - y = -2 \end{cases}$$

$x - y = -2$ will become:

$x = -2 + y$

We substitute the expression of x into the first equation.

$3(y - 2) + 4y = -5$

$3y - 6 + 4y = -5$

$7y - 6 = -5$

$7y = -5 + 6$

$7y = 1$

$y = \frac{1}{7} = 0.14$

We substitute the value of y into the second equation.

$x - 0.14 = -2$

$x = -2 + 0.14 = -1.86$

3) Find the solution of the system of equations

$$\begin{cases} 3x + 3y = 4 \\ x - 2y = 3 \end{cases}$$

We isolate x in the second equation

$x = 2y + 3$

We substitute this expression for x in the first equation and solve for y

$3(2y + 3) + 3y = 4$

$6y + 9 + 3y = 4$

$9y + 9 = 4$

$9y = 4 - 9$

$9y = -5$

$y = -\dfrac{5}{9} = -0.55$

We substitute the value of y in the second equation to find the value of x.

$x - 2y = 3$

$x - 2(-0.55) = 3$

$x + 1.1 = 3$

$x = 3 - 1.1$

$x = 1.9$

4) Find the solution of the system of equations

$$\begin{cases} 3x + 2y = 1 \\ 3x - y = 2 \end{cases}$$

We isolate y in the second equation

$3x - 2 = y$

We substitute this expression for y in the first equation and solve for x

$3x + 2(3x - 2) = 1$

$3x + 6x - 4 = 1$

$9x - 4 = 1$

$9x = 1 + 4$

$9x = 5$

$x = \dfrac{5}{9} = 0.55$

We substitute the value of x in the second equation to find the value of x.

$3(0.55) + 2y = 1$

$1.66 + 2y = 1$

$2y = 1 - 1.66$

$2y = -0.66$

$y = -\dfrac{0.66}{2} = -0.33$

5) Find the solution of the system of equations

$$\begin{cases} \dfrac{y+3}{1-2x} = \dfrac{1}{4} \\ x + 3y = 2 \end{cases}$$

We isolate x in the second equation

$x = 2 - 3y$

We substitute this expression for x in the first equation and solve for y

$\dfrac{y+3}{1-2(2-3y)} = \dfrac{1}{4}$

$\dfrac{y+3}{1-4+6y} = \dfrac{1}{4}$

$\dfrac{y+3}{-3+6y} = \dfrac{1}{4}$

We are using the cross-multiplication property for fractions

$4(y + 3) = 6y - 3$

$4y + 12 = 6y - 3$

$12 + 3 = 6y - 4y$

$15 = 2y$

$2y = 15$

$y = \frac{15}{2} = 7.5$

We substitute the value of y in the second equation to find the value of x.

$x + 3(7.5) = 2$

$x + 22.5 = 2$

$x = 2 - 22.5 = -20.5$

c. Solving a system of linear equations by Method of Elimination

1) Solve the system of equations using elimination:
$$\begin{cases} y = 5x - 6 \\ y = 3x + 4 \end{cases}$$

If we subtract vertically, we have:

$0 = 2x - 10$

$2x = 10$

$x = 5$

$y = 5(5) - 6 = 25 - 6 = 19$

2) Solve the system of equations using elimination:
$$\begin{cases} 2x + 3y = 4 \\ x - 2y = 3 \end{cases}$$

We multiply the second equation by minus 2

$$\begin{cases} 2x + 3y = 4 \\ -2x + 4y = -6 \end{cases}$$

Add vertically

$7y = -2$

Divide by 7

$y = -\frac{2}{7} = -0.28$

So, we substitute y into the second original equation

$x - 2(-0.28) = 3$

$x = -2(0.28) + 3$

$x = 2.42$

3) Solve the system of equations using elimination:

$$\begin{cases} \frac{y+2}{1-2x} = \frac{1}{3} \\ x + 3y = 2 \end{cases}$$

In the first equation, we are using the cross-multiplication property

$$\begin{cases} 3(y + 2) = 1 - 2x \\ x + 3y = 2 \end{cases}$$

$$\begin{cases} 3y + 6 = 1 - 2x \\ x + 3y = 2 \end{cases}$$

$$\begin{cases} 3y + 2x + 6 - 6 = 1 - 6 \\ x + 3y = 2 \end{cases}$$

$$\begin{cases} 2x + 3y = -5 \\ x + 3y = 2 \end{cases}$$

We multiply the second equation with minus one

$$\begin{cases} 2x + 3y = -5 \\ -x - 3y = -2 \end{cases}$$

Add vertically

$x = -7$

In first equation

$3(-7) + y = 5$

$-21 + y = 5$

$y = 5 + 21 = 26$

4) Solve the system of equations using elimination:

$$\begin{cases} \frac{5}{2}x - \frac{2}{3}y = 11 \\ 5x - 3y = 2 \end{cases}$$

We multiply the first equation by 6 (common denominator.

$$\begin{cases} \frac{5*6}{2}x - \frac{2*6}{3}y = 66 \\ 5x - 3y = 2 \end{cases}$$

So,

$$\begin{cases} 15x - 4y = 66 \\ 5x - 3y = 2 \end{cases}$$

We multiply the second equation by minus 3

$$\begin{cases} 15x - 4y = 66 \\ -15x + 9y = -6 \end{cases}$$

Add vertically

$5y = 60$

$y = \frac{60}{5} = 12$

We substitute y into the second ORIGINAL equation

$5x - 3(12) = 2$

$5x = 2 + 36$

$5x = 38$

$x = \frac{38}{5} = 7.6$

5) Solve the system of equations using elimination:

$$\begin{cases} \frac{1}{2}x + \frac{2}{5}y = 1 \\ \frac{5}{6}x - \frac{1}{2}y = 2 \end{cases}$$

We multiply the first equation by 10 and the second by 6

$$\begin{cases} \frac{10}{2}x + \frac{2*10}{5}y = 10 \\ \frac{5*6}{6}x - \frac{1*6}{2}y = 12 \end{cases}$$

So,

$$\begin{cases} 5x + 4y = 10 \\ 5x - 3y = 12 \end{cases}$$

We multiply the second equation by minus 1

$$\begin{cases} 5x + 4y = 10 \\ -5x + 3y = -12 \end{cases}$$

Add vertically

$$7y = -2$$

$$y = -\frac{2}{7} = -0.28$$

We substitute y into the first ORIGINAL equation

$$\frac{1}{2}x + \frac{2}{5}(-0.28) = 1$$

$$0.5x - 0.11 = 1$$

$$0.5x = 1 + 0.11$$

$$0.5x = 1.11$$

$$x = \frac{1.11}{0.5} = 2.22$$

ABOUT THE AUTHOR

Dr. Marcel Sincraian has been working with numbers whole his life as an Engineer, Accountant, Math and Physics teacher. While an Engineer, he got his Ph.D. in Civil Engineering. He participated in European engineering research projects in soil dynamics. He published papers in international Engineering Journals and International Conferences. He is a passionate, warm, and funny Mathematics and Physics teacher. He is always ready to share his knowledge in Math and Physics. Because of his teaching passion, he decided to help the students with a few math books, from Fundamentals of Algebra to Introduction to Calculus to pure Calculus. He is continually writing new Math and Physics books hoping to help students with other parts of Mathematics and Physics. His hobbies are, reading, traveling, camping.